ELECTRICAL MACHINES FUNDAMENTALS

BY PRASUN BARUA

ABOUT

"Electrical Machines Fundamentals" is a comprehensive book offering a thorough understanding of electrical machinery, from transformers to generators and motors. Beginning with single-phase transformers, it delves into construction, losses, and testing procedures before expanding to encompass three-phase transformers and auto transformers. With detailed discussions on D.C. generators and motors, including speed control and starting mechanisms, and in-depth coverage of three-phase induction motors, including rotor types and performance evaluation tests, the book bridges theoretical understanding with practical application. It explains concepts such as regulation, efficiency, and advanced topics like electromechanical energy conversion principles and emerging technologies like AC servo motors and dielectric heating. The book provides insights essential for designing, operating, and optimizing electrical systems in contemporary power networks. Thanks for reading the book.

TABLE OF CONTENTS

CHAPTER	TOPIC	PAGE
1	Single Phase Transformer	4
1.1	Construction of Single-Phase Transformer	7
1.2	Losses in Transformer	11
1.3	Open-Circuit and Short-Circuit Test	15
2	Three-Phase Transformer	19
3	Auto Transformer	27
4	D.C. Generators	31
4.1	Speed Control of D.C. Motors	34
4.2	Starting of D.C. Motors	39
4.3	Types of D.C Machine	42
5	Three Phase Induction Motors	49
5.1	Types of Three Phase Induction Motor Rotor	52
5.2	No-Load and Blocked Rotor Test	55
5.3	Torque-Slip and Torque-Speed Characteristics	58
5.4	Equivalent Circuit of an Induction Motor	61
6	Parallel Operations of Transformers	65
7	Regulation and Efficiency	69
8	Types of Single-Phase Induction Motors	74
8.1	Working Principle of a Single-Phase Induction Motor	80
9	Construction of Three Phase Synchronous Machines	83
10	Electromechanical Energy Conversion Principles	86
11	Armature Reaction in DC Generator	88
12	Armature Reaction in Synchronous Generator	96
13	Armature Reaction in DC Machine	103
14	What is an AC Servo Motor	108
15	Applications of Three-Phase Induction Motor	116
16	What is Dielectric Heating	120

CHAPTER-1: SINGLE PHASE TRANSFORMER

The transformer is a static device which converts the magnetic energy into electrical energy. It consists of two or more than two stationary circuits interlinked by a common magnetic circuit; the energy transfer takes place through this circuit without having any change in frequency from one circuit to another.

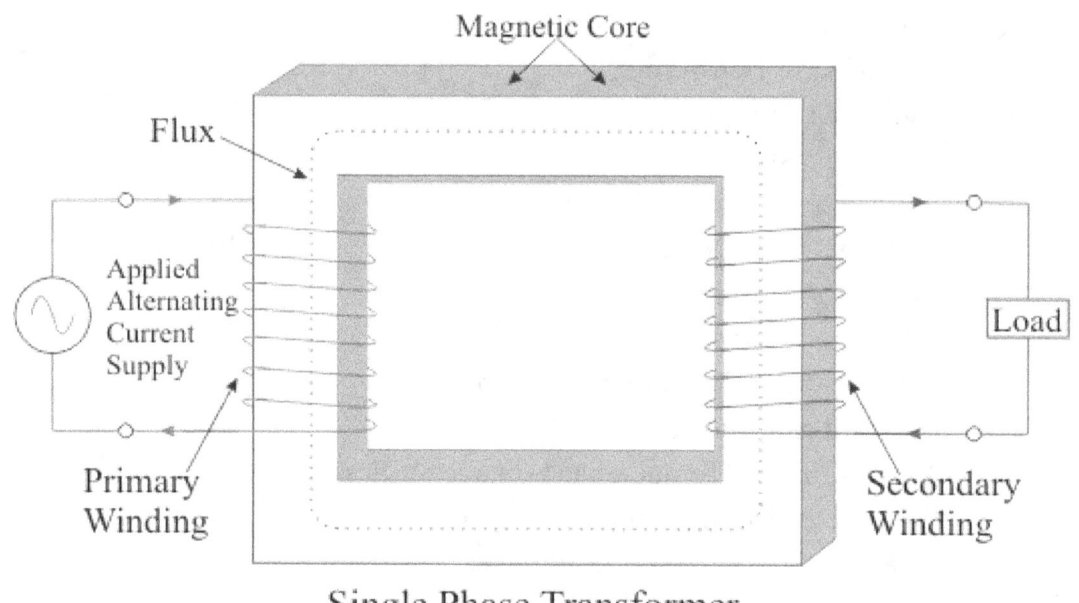

Single Phase Transformer

The transformer consists of two windings. The winding which is connected with the supply a.c. voltage is called the primary winding, and the winding which is connected to load and delivers the energy to load is called secondary winding.

E.M.F. Equation of a Transformer

Let the flux at any point is given by

$\varphi = \varphi_m \sin \omega t$

The instantaneous e.m.f. induced in a coil of T turns linked by this flux is given by Faraday's law as

$$e = -\frac{d}{dt}(\varphi T) = -T\frac{d\varphi}{dt} = -T\frac{d}{dt}(\varphi_m \sin \omega t) = -T\omega\varphi_m \cos \omega t$$

$$T\omega\varphi_m \sin\left(\omega t - \frac{\pi}{2}\right)$$

The above equation can also be written as

$$e = E_m \sin\left(\omega t - \frac{\pi}{2}\right)$$

Where $E_m = T\omega\varphi_m$ = maximum value of e.

For sine wave, the r.m.s. value is given by

$$E_{rms} = E = E_m/\sqrt{2}$$

$$E = \frac{T\omega\varphi_m}{2} = \frac{T(2\pi f)\varphi_m}{\sqrt{2}}$$

$E_{rms} = E = E_m/$ $\boxed{E = 4.44\varphi_m fT}$

This is called as e.m.f. equation of transformer.

Where,

φ_m is the maximum flux in webers (Wb)

f is the frequency in hertz (Hz)

E is the voltage in volts

T is number of turns in winding

The primary r.m.s. voltage is

$E_1 = 4.44\varphi_m fT_1$

The secondary r.m.s. voltage is

$E_2 = 4.44 \varphi_m f T_2$

Voltage Ratio and Turns Ratio

The ratio E/T is called *voltage per turn*.

As we know

$E_1 = 4.44 \varphi_m f T_1$ therefore

$$\frac{E_1}{T_2} = 4.44 \varphi_m f \qquad \ldots\ldots\ldots\text{eq.1}$$

And $E_2 = 4.44 \varphi_m f T_2$ therefore

$$\frac{E_2}{T_2} = 4.44 \varphi_m f \qquad \ldots\ldots\ldots\text{eq.2}$$

From equation 1 and 2

$$\frac{E_1}{T_1} = \frac{E_2}{T_2}$$

Also, $\dfrac{E_1}{E_2} = \dfrac{T_1}{T_2}$

The ratio $\dfrac{T_1}{T_2}$ is called **turns ratio.**

Step - up transformer: These are the transformer in which the output voltage is higher than the input voltage.

Step - down transformer: These are the transformer in which the output voltage is less than the input voltage.

NOTE: The same transformer can be used as a step up transformer and step down transformer by changing the way it is connected. If we want the transformer to work as a step up transformer, then low voltage winding is the primary, and if we want it to work as a step-down transformer, then high voltage winding is the primary.

1.1 Construction of Single - Phase Transformers

A single-phase transformer comprises primary and secondary windings wrapped around a core. This core is crafted from thin sheets, known as laminations, made of high-grade silicon steel. These laminations serve a dual purpose: they minimize eddy-current loss, and the silicon steel helps in reducing hysteresis loss. To ensure insulation between laminations, a heat-resistant enamel coating is applied. Two common types of laminations used in transformer construction are L-Type and E-Type. There are two basic types of transformer constructions:

1. Core type construction.
2. Shell type construction.

Core type Construction

In a core-type transformer, the magnetic circuit comprises two vertical legs or limbs along with two horizontal sections known as yokes. To diminish the leakage flux, half of each winding is positioned on each leg of the core. The low voltage winding is positioned adjacent to the core, while the high voltage winding wraps around the low voltage winding, reducing the amount of insulating material needed. This arrangement forms concentric coils, commonly referred to as concentric or cylindrical winding.

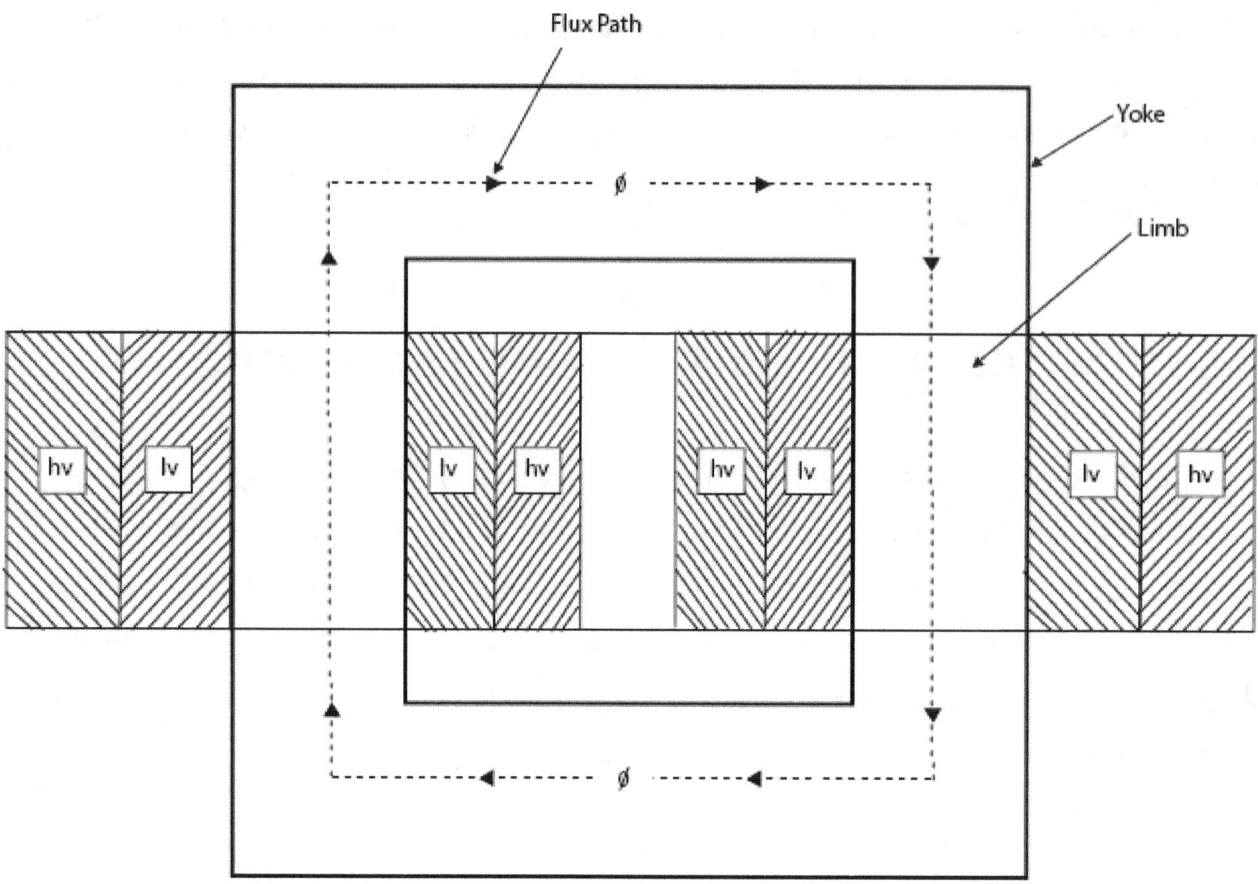

Shell type Construction

In the shell type transformer, both the primary and secondary winding are wounded on the central limb, and the low reluctance path is completed by the outer limbs. Each winding is subdivided into sections. Low voltage (lv) and High voltage (hv) subsections are alternatively placed in the form of sandwich that is why this winding is also called sandwich or disc winding.

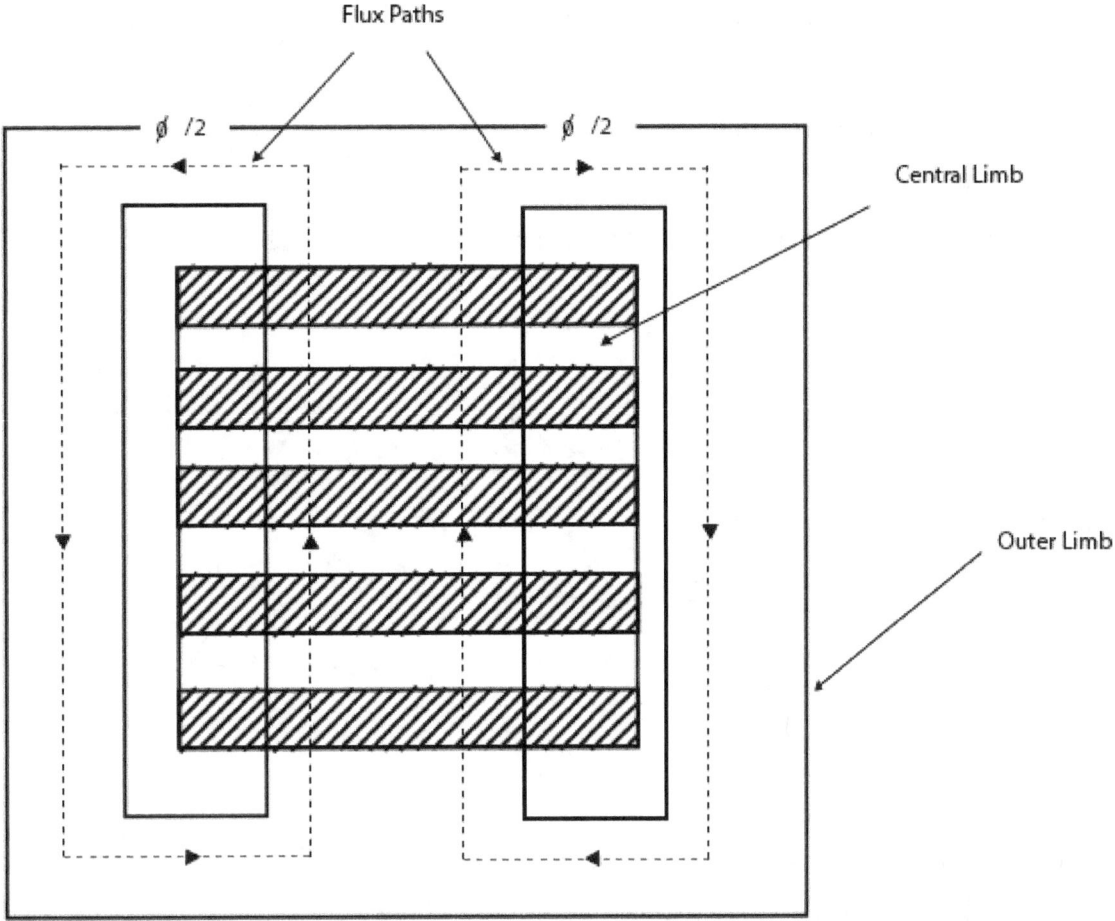

The core comprises two types of laminations: U-shaped and I-shaped. Initially, the U-shaped laminations are stacked together to achieve the desired length. Half of the prewound low voltage coil is then wrapped around the limbs of the core, followed by insulation. Next, half of the prewound high voltage coil is placed around the low voltage coil. Finally, the core is sealed shut using the I-shaped laminations at the top.

Ideal Transformer

An Ideal transformer is an imaginary transformer which has the following properties:

1. Its primary and secondary winding resistances are negligible.

2. The core has infinite permeability (μ) so that negligible mmf is required to establish the flux in the core.
3. Its leakage flux and leakage inductances are zero. The entire flux is confined to the core and links both windings.
4. There are no losses due to resistance, hysteresis and eddy currents. Thus, the efficiency is 100 percent.

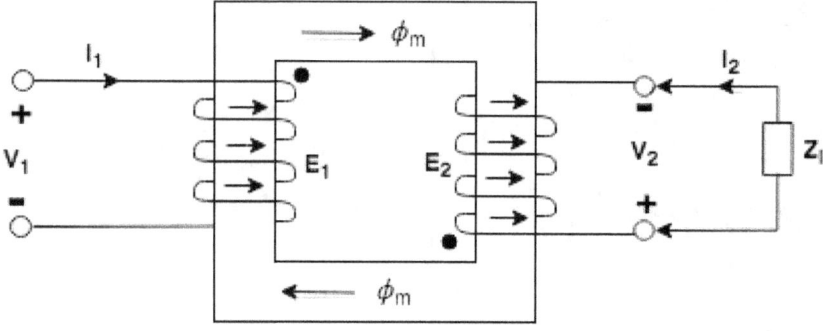

Figure: Ideal iron-core transformer

The transformer that has zero primary and zero secondary impedance, is called an ideal transformer. The applied voltage V1 in the primary is equal to the induced voltage E1. Similarly, the induced voltage E2 is equal to the output voltage V2 of the secondary.

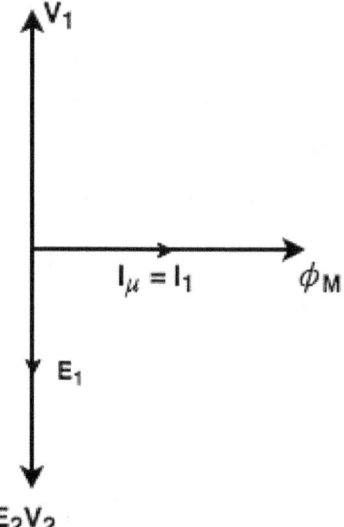

Figure: No-load Phasor diagram of an ideal transformer.

For an ideal transformer, if a = transformation ratio = turn ratio

Then,

$$a = \frac{T_1}{T_2} = \frac{E_1}{E_2} = \frac{V_1}{V_2} = \frac{I_2}{I_1} \quad \text{eq. 1}$$

$$\therefore I_1 T_1 = I_2 T_2 \quad \text{eq. 2}$$

$$E_1 I_1 = E_2 I_2 = S_2 = S_1 \quad \text{eq. 3}$$

$$V_1 I_1 = V_2 I_2 = S_2 = S_1 \quad \text{eq. 4}$$

The equation-2 states that the demagnetizing, ampere-turns of the secondary are equal and opposite to the magnetizing mmf of the primary of an Ideal transformer.

Applications of Transformers

- The level of voltage and current can be changed in electrical power systems.
- Transformer known as instrument transformer is used to measure the voltage and current.
- In combined ac/dc power systems, the transformers are used to convert hvac to hvdc.
- To isolate one circuit from another, since primary and secondary are not connected.

1.2 Losses in Transformers

There are two types of losses occurs in a transformer:

1. Iron loss or Core loss **P_i**
2. Copper loss or **I^2R** loss **P_c**

Iron loss or core loss (P_i)

Iron loss in transformers is the combination of hysteresis loss (P_h) and eddy current loss (P_e). This type of loss mainly occurs in the magnetic core of the transformer, and depends on magnetic properties of core material.

$$P_i = P_h + P_e$$

The formula for hysteresis and eddy current losses is as follows:

$$P_h = k_h f B_m^x$$

$$P_e = k_e f^2 B_m^2$$

Where,

k_h = It is a constant which is proportional to the volume, quantity of the core material and the units used.

k_e = It is a constant which is proportional to the volume, resistivity of the core material, thickness of laminations and the units used.

B_m = maximum flux density in the core.

f = frequency of the alternating flux.

The exponent 'x' is called Steinmetz constant. Depending upon the magnetic properties of the core material its value varies from 1.5 to 2.5.

Therefore the total core loss in the transformer is

$$P_i = P_h + P_e$$

$$P_i = k_h f B_m^x + k_e f^2 B_m^2$$

As we know that the voltage applied is approximately equal to the induced voltage in the transformer.

$V_1 = E_1 = 4.44\, \Phi_m f T_1 = 4.44\, B_m A_i f T_1$

$B_m = \dfrac{V_1}{4.44\, A_i f T_1}$

And
$$P_h = k_h f B_m^x = k_h f \left(\dfrac{V_1}{4.44\, A_i f T_1}\right)^x$$

$$= k_h \left(\dfrac{1}{4.44\, A_i f T_1}\right)^x \cdot f\left(\dfrac{V_1}{f}\right)^x = = K_h f\, V_1^x f^{1-x}$$

Where
$$K_h = k_h \left(\dfrac{1}{4.44\, A_i T_1}\right)^x$$

The above relation shows that the hysteresis loss depends upon both the applied voltage and frequency.

$$P_e = k_e f^2 B_m^2 = k_e f^2 \left(\dfrac{V_1}{4.44\, A_i T_1 f}\right)^2 = k_e V_1^2$$

The above relation shows that the eddy current loss is proportional to the square of the applied voltage and is independent of frequency.

Since, $V_1 = 4.44\, B_m A_i f T_i$,

$V_1 \propto B_m f$

Which means for any voltage if f decreases, B_m increases. Similarly, if f increases, B_m decreases.

The total core loss can be written as

$P_i = K_h V_x f_{1-x} + K_e V^2$

Copper loss or I²R loss (P_c)

The loss which takes place in the primary and secondary winding of the transformer because of the winding resistance is called the Copper loss or I²R loss.

Total copper loss in the transformer = Primary winding copper loss + Secondary winding copper loss

$$P_c = I_1^2 R_1 + I_2^2 R_2$$

Since $I_1 T_1 = I_2 T_2$

$$I_1 = I_2 \frac{T_2}{T_1}$$

$$\therefore P_c = I_2^2 \left(\frac{T_2}{T_1}\right) R_1 + I_2^2 R_2 = I_2^2 \left[R_2 + \left(\frac{T_2}{T_1}\right)^2 R_1\right] = I_2^2 R_{e_2}$$

Therefore copper loss varies as the square of the load current.

$$P_c = I_1^2 R_1 + I_2^2 R_2 = I_1^2 R_1 + \left(I_1 \frac{T_1}{T_2}\right)^2 R_2 = I_1^2 \left[R_1 + \left(\frac{T_1}{T_2}\right)^2 R_2\right] = I_1^2 R_{e_2}$$

$$\therefore P_c = I_1^2 R_{e_1} = I_2^2 R_{e_2}$$

Stray Loss

Eddy current in the conductor, tank, etc., produced by the leakage flux in a transformer is known as stray losses. We can neglect these losses, as the percentage of these losses is very less as compared to iron and copper loss.

Dielectric loss

The losses that occur in insulating materials, i.e., in the transformer oil and the solid insulation of the transformer, are known as a dielectric loss. This loss occurs only in high voltage transformer, and is very small so that we can neglect these losses.

1.3 Open Circuit Test and Short Circuit Tests

To determine the circuit constant, efficiency and regulation of a transformer, without actually loading the transformer, we perform Open-Circuit and Short-Circuit tests. These tests give more accurate result than compared with the fully loaded transformer. And the power consumption in these tests is minimal as compared with the transformer's output on full load.

Open Circuit Test

The circuit diagram of the open circuit test for the transformer is shown below:

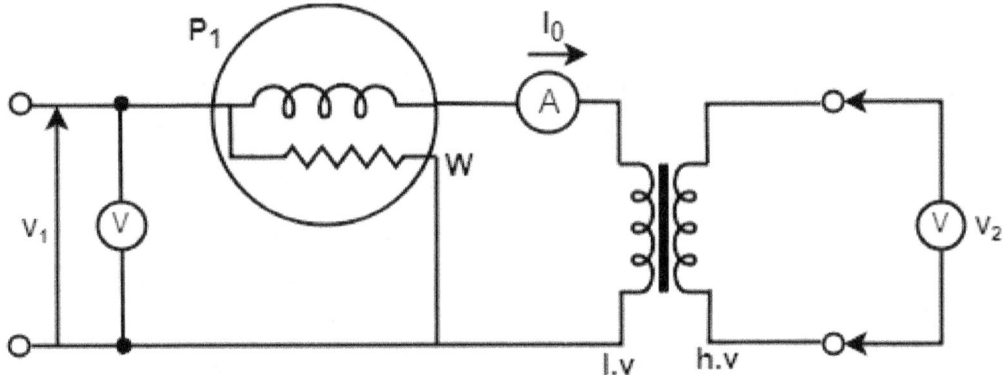

Fig: Open Circuit Test on a Transformer

A voltmeter V, an ammeter A, and a wattmeter W are connected in the low voltage (lv) side of transformer. The voltmeter v gives the rated voltage V_1 of the primary. A very small current I_0, called the no-load current, flows in the primary side because the secondary side is open circuited. The ammeter A, therefore, reads the no-load current I0. The power loss in the transformer occurs due to core loss and a very small I_2R loss in the primary. There is no I2R loss in the secondary since it is open and $I_2 = 0$. Since the no-load current I_0 is very small, the I_2R loss in the primary winding can be neglected. The instrument readings obtained in open circuit test are as follows:

- Ammeter reading = no-load current I_0
- Voltmeter reading = primary rated voltage V_1
- Wattmeter reading = iron or core loss P_i

From the above readings we can determine the no-load equivalent circuit:

$P_i = V_1 I_0 \cos\Phi_0$

No-load power factor, $\cos\Phi_0 = \dfrac{P_i}{V_1 I_0}$

$I_w = I_0 \cos\Phi_0$

$I_\mu = I_0 \sin\Phi_0$

$R_0 = \dfrac{V_1}{I_w}$, $X_0 = \dfrac{V_1}{I_\mu}$

Where X_0 is the inductive reactance, I_w is the core loss current, I_μ is the magnetizing current.

Short Circuit Test

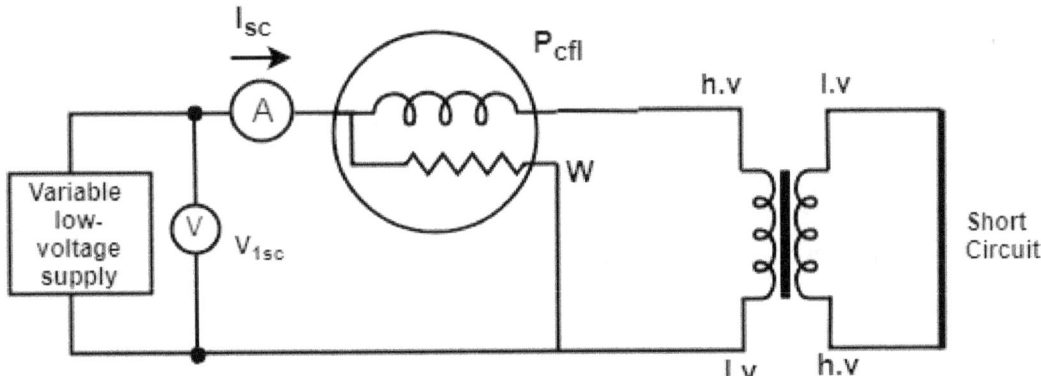

Fig: Short Circuit Test on a Transformer

In the short circuit test, usually, the low voltage side is short-circuited by a thick conductor. As we can see that in the figure an ammeter, a voltmeter, and a wattmeter are connected on the high-voltage side. The reason for short-circuiting the low voltage side is as follows:

1. The rated current on the high voltage (hv) side of a transformer is less than the low voltage (lv) side of the transformer. And we can measure this current with the help of available laboratory ammeters.
2. Greater accuracy in the reading of the voltmeter is possible when we use the hv side as the primary because the applied voltage is less than 5 percent of the rated voltage of the winding.

The high voltage winding is supplied at the reduced voltage from a variable voltage supply. This supply voltage is gradually increased until full-load primary current flows. When the rated full load current flows in the primary winding, then it will also flow in the secondary winding by the transformer action.

The readings of the instruments on the short circuit test are as follows:

- Ammeter reading = Full load primary current, I_{1SC}
- Voltmeter reading = Short circuit voltage, V_{1SC}

- Wattmeter reading = full-load copper loss of the transformer P_{cfl}

The equivalent resistance of the transformer referred to primary

$$R_{e_1} = \frac{P_{cfl}}{I_{1SC}^2}$$

Equivalent impedance referred to primary

$$Z_{e_1} = \frac{V_{1SC}}{I_{1SC}}$$

Equivalent reactance referred to primary

$$X_{e_1} = \sqrt{Z_{e_1}^2 - R_{e_1}^2}$$

$$\cos \Phi_{sc} = \frac{R_{e_1}}{Z_{e_1}}$$

When short circuit test is performed only on one side the equivalent circuit constants referred to another side can also be calculated as follows:

$$Z_{e_2} = Z_{e_1}\left(\frac{T_2}{T_1}\right)^2 = \frac{Z_{e_1}}{a^2}$$

$$R_{e_2} = R_{e_1}\left(\frac{T_2}{T_1}\right)^2 = \frac{R_{e_1}}{a^2}$$

$$X_{e_2} = X_{e_1}\left(\frac{T_2}{T_1}\right)^2 = \frac{X_{e_1}}{a^2}$$

CHAPTER-2: THREE PHASE TRANSFORMERS

The three-phase transformers are mainly of two types:

1. Single unit three-phase transformer.
2. Three unit three-phase transformer.

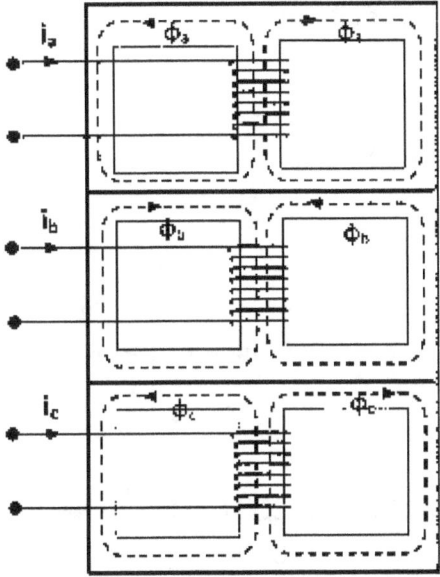

3-Phase Shell Type Transformer

Figure: Three phase shell type transformer

Advantages of Bank of three units:

In case of damage of one winding the power can be transmitted through the two units by using an open delta connection, so 50% power can be transferred.

Advantages of single unit transformer:

1. These transformers use less space.
2. These transformers are lighter, smaller, and cheaper.
3. These transformers are slightly more efficient.

Disadvantages of single unit transformer:

If the single winding of the transformer gets damaged, then we have to change the complete unit.

Connections of three-phase transformer:

A three-phase transformer has three transformers connected in it, either separately or combined on one core. We can connect the primary and secondary winding of a 3-phase transformer in either a star (Y) or delta (Δ). There are four ways to connect the 3-phase transformer bank:

1. Δ - Δ (Delta primary - Delta secondary)
2. Y - Y (Star primary - Star secondary)
3. Δ - Y (Delta primary - Star secondary)
4. Y - Δ (Star primary - Delta secondary)

For connecting the transformers into star or delta, we have to assume that the transformers we are connecting, are all of the same KVA ratings.

What are the factors that affect the choice of connections?

The factors that affect the choice are as follows:

1. We have to check the availability of a neutral connection for grounding, protection or load current.
2. Insulation to ground and voltage stress.
3. We have to check that the path for the flow of third harmonic and zero sequence current is available or not.
4. When one circuit is out of service, we need the partial capacity.
5. Parallel operation with other transformers.
6. We have to check the economic considerations.

1. Delta - Delta (Δ - Δ) Connection

In delta-delta connection the line voltage of the transformer is equal to the supply voltage of the transformer.

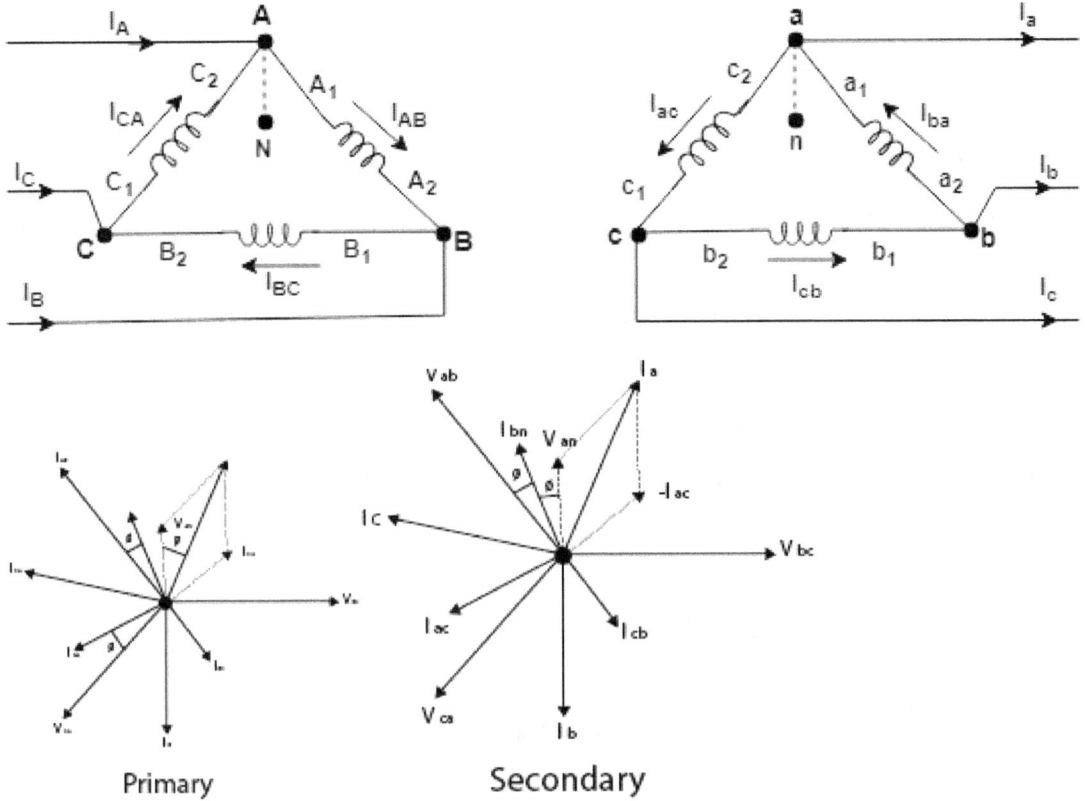

Primary   Secondary

The above diagram shows the delta-delta connection of three windings of single phase transformer. The secondary winding $a_1 a_2$ corresponds to the primary winding $A_1 A_2$, $b_1 b_2$ corresponds to $B_1 B_2$, and $c_1 c_2$ corresponds to $C_1 C_2$, similarly 'a' corresponds to A, 'b' corresponds to B and 'c' corresponds to C. The terminals 'a_1' and A_1 have the same polarity. The Phasor diagram drawn above is for lagging power factor cos Φ. For balanced conditions, the line current is $\sqrt{}$ three times the Phase current.

The turn ratio for 3 phase transformer is

$$\frac{V_{AB}}{V_{ab}} = \frac{V_{BC}}{V_{bc}} = \frac{V_{CA}}{V_{ca}} = a$$

And the current ratio when the magnetizing current is neglected is

$$\frac{I_{AB}}{I_{ab}} = \frac{I_{BC}}{I_{bc}} = \frac{I_{CA}}{I_{ca}} = \frac{I_A}{I_a} = \frac{I_B}{I_b} = \frac{I_C}{I_c} = \frac{1}{a}$$

We can understand from the above-drawn diagram that both the primary and secondary line voltages are in phase. This connection is called **0° - connection**.

If we reverse the connections of the phase winding, we get the phase difference of 180° between the primary and secondary systems. This connection is called **180° - connection**.

The advantage of Δ - Δ Transformation:

1. The delta-delta connection is good for balanced and unbalanced loading.
2. If a third harmonic is present, it circulates in the closed path and therefore does not appear in the output voltage wave.
3. The main advantage of Δ - Δ transformer is that if one transformer stops working, then the other two transformers will keep on working. This is called an open delta connection.

The disadvantage of Δ - Δ Transformation:

The disadvantage of Δ - Δ transformer is that it does not contain a neutral point and this can only be used when neither primary nor secondary requires neutral, and the required voltage is low and moderate.

2. Star-Star (Y - Y) Connection

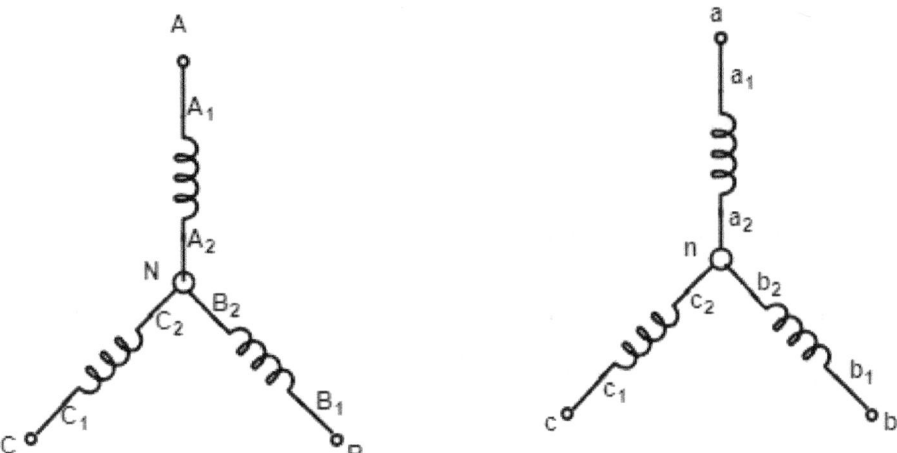

Fig: Star-Star connection of transformer (0° Phase Shift)

The voltage ratios for ideal transformer are:

$$\frac{V_{AN}}{V_{an}} = \frac{V_{BN}}{V_{bn}} = \frac{V_{CN}}{V_{cn}} = a$$

And current ratios are:

$$\frac{I_A}{I_B} = \frac{I_B}{I_b} = \frac{I_C}{I_c} = \frac{1}{a}$$

There are two serious problems in star-star connection:

1. In the star-star connection when the load is unbalanced and neutral is not provided, then the phase voltage tends to become severely unbalanced. Therefore, the star-star connection is not suitable for unbalanced loading.

2. The magnetizing current of any transformer is very non-sinusoidal and contains a very large third harmonic, which is necessary to overcome saturation to produce a sinusoidal flux.

3. Delta-Star (Δ - Y) Connection

In Δ - Y connection of 3-phase transformers, the primary line voltage is equal to the primary phase voltage ($V_{LP} = V_{pP}$). The relationship between secondary voltages is $V_{LS} = \sqrt{3}\, V_{pS}$ therefore, the line-line voltage ratio of this connection is

$$\frac{V_{LP}}{V_{LS}} = \frac{V_{pP}}{\sqrt{3}\, V_{pS}}$$

$$\frac{V_{pP}}{V_{pS}} = a$$

$$\therefore \frac{V_{LP}}{V_{LS}} = \frac{a}{\sqrt{3}}$$

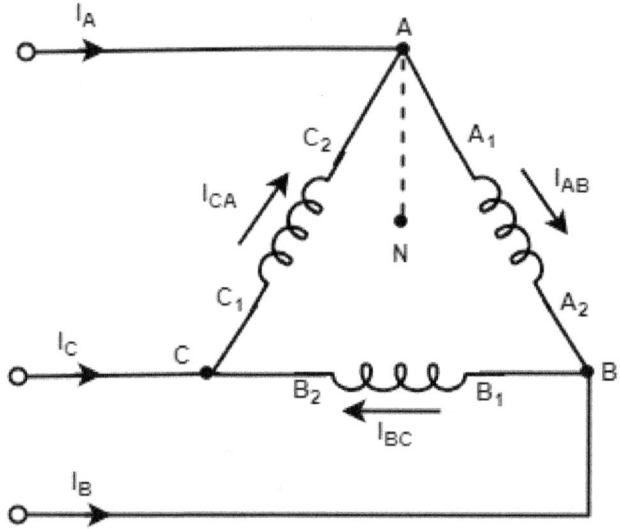
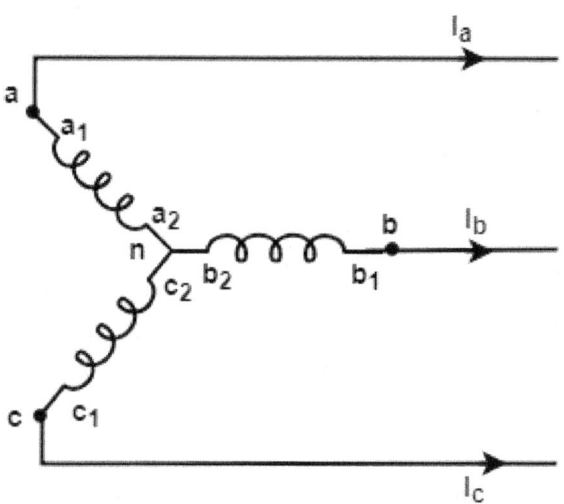

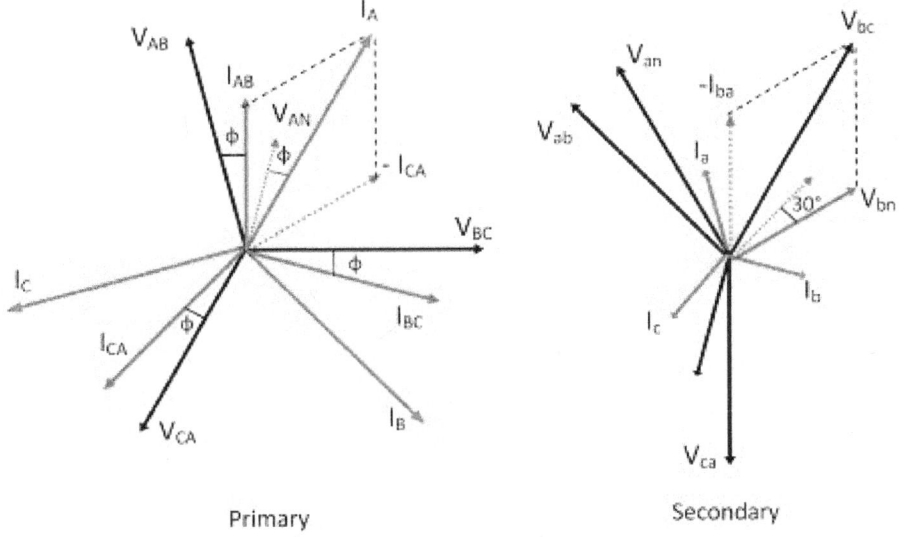

Primary Secondary

Phasor Diagram of Delta-Star Connection of Transformer

(b)

Figure: Delta - star connection of transformer (Phase shift 30° lead), (b) Phasor diagram

The phasor diagram drawn above shows the delta-star connection supplying a balanced load at power factor cos Φ lagging. It is seen from the phasor diagram that the secondary phase voltage V_{an} leads primary phase voltage V_{AN} by 30°. Similarly, V_{bn} leads V_{BN} by 30° and V_{cn} leads V_{CN} by 30°. This is also the phase relationship between the respective line-to-line voltages. This connection is called **+30° connection**.

4. Star-Delta (Y - Δ) Connection

The **Y - Δ** connection of three-phase transformers is shown below. In this connection, the primary line voltage is equal to $\sqrt{3}$ times the primary phase voltage ($V_{LP} = \sqrt{3} V_{pP}$). The secondary line voltage is equal to the secondary phase voltage ($V_{LS} = V_{pS}$). The voltage ration of each phase is

$$\frac{V_{pP}}{V_{pS}} = a$$

25

Therefore line-to-line voltage ratio of a **Y - Δ** connection is

$$\frac{V_{LP}}{V_{LS}} = \frac{\sqrt{3}V_{pP}}{V_{pS}} = \sqrt{3}\,a$$

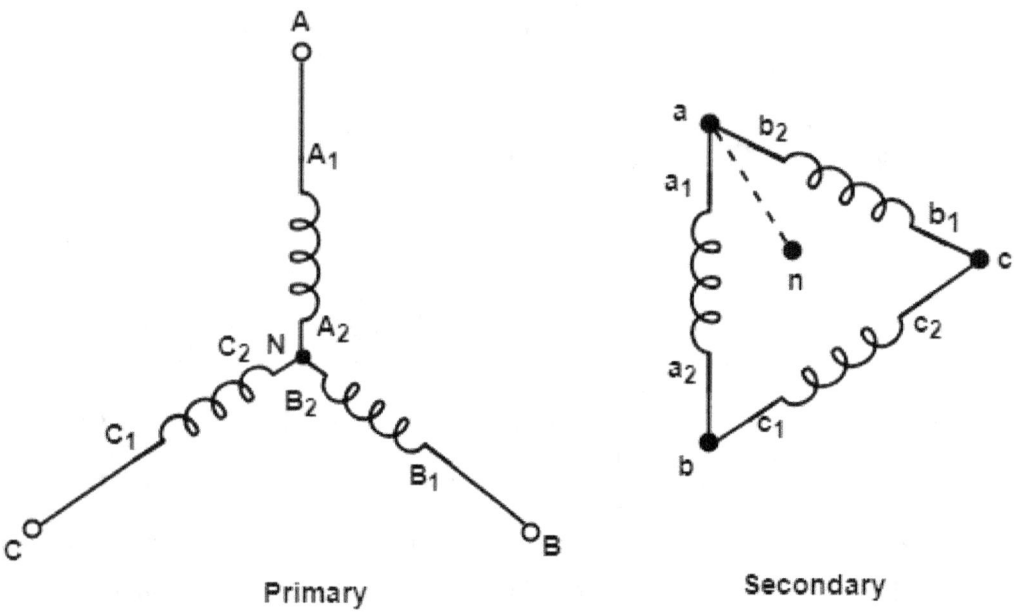

Figure: Y - Δ connection of transformer (Phase shift of 30° lead)

When there is a phase shift of 30° lead between respective line-to-line voltages then this type of connection is called as **+30° connection**.

And when there is a phase shift of 30° lag between the line-to-line voltages then the connection is known as **-30° connection**.

The Δ - Y connection or Y - Δ connection has no problem with unbalanced loads and third harmonics.

CHAPTER-3: AUTO TRANSFORMER

What is single-phase autotransformer?

It is a single winding transformer in which a part of the winding is common to both high-voltage and low - voltage sides

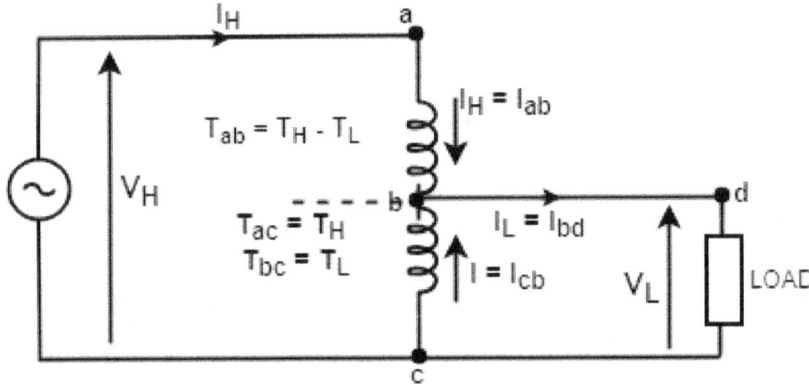

Fig: Step-down autotransformer

In the figure drawn above 'abc' is a single winding in which terminal 'a' and 'c' is the high voltage terminal and 'b' and 'c' is the low voltage terminal. The winding 'b' and 'c' is known as the common winding and winding 'a' and 'b' is known as the series winding.

A step-down transformer is a transformer in which the primary voltage is higher than the secondary voltage. Here in the above diagram voltage 'V_H' is applied to full winding 'abc' and the load is connected to winding 'bc'. This arrangement is called as Step down Autotransformer.

In an autotransformer, the windings are physically connected that is why a different type of terminology is used for this transformer.

$T_H = T_{ac}$ = number of turns in full winding 'abc'

= number of turns in hv side

$T_L = T_{bc}$ = number of turns of the common winding 'bc'

= number of turns of the lv side

$T_{ab} = T_H - T_L$ = number of turns of the series winding 'ab'

V_H = input voltage on the hv side

V_L = output voltage on the lv side

I_H = input current in the hv side

I_L = output current in the lv side

Current in the series winding = I_{ab} = IH

Current in the common winding 'bc' = I_{cb} = I

Step up autotransformer

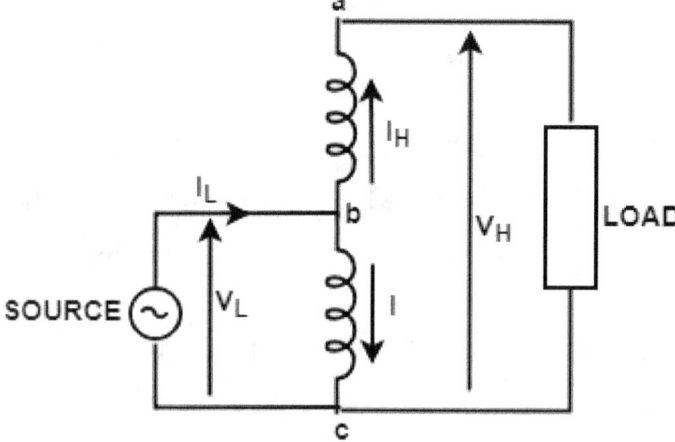

Fig: Step-up autotransformer

The step-up transformer is shown in the above figure, and we can say that the source is connected to the winding 'bc' and load is connected with the complete winding 'abc'.

Advantages of Autotransformers

1. In an autotransformer, less winding material is used as compared with two winding transformers.
2. Autotransformers are smaller in size.
3. These are cheaper than the two winding transformers with the same output.
4. Because of less use of conductor and core materials, the ohmic losses in the conductor and the core losses are smaller.
5. Autotransformers have high efficiency as compared with two winding transformers.
6. An autotransformer has variable output voltage when a sliding contact is used for the secondary.

Disadvantages of Autotransformers

1. In an autotransformer, the effective per unit impedance is smaller than 2 - winding transformer.
2. In case of open circuit winding, full voltage is transferred from low voltage side to high voltage side this high voltage can burn out or damage the equipment connected on its secondary side.
3. In an autotransformer, there is a loss of isolation between input and output circuits.

Applications of Autotransformers

1. Autotransformers are used for interconnecting power systems of different voltage levels like 132KV and 230KV.
2. These are used to boost the supply voltage by a small amount in distribution systems for the compensation of voltage drop.

3. If an autotransformer has a number of tappings than it can be used for starting induction motors and synchronous motors.
4. These transformers can be used as variac i.e. variable a.c. in the laboratory and if a continuously variable voltage is required over broad ranges.

CHAPTER-4: D.C. GENERATORS

Basic Structure of Electrical machines:

A rotating machine comprises two essential components: the stator and the rotor. The stator remains stationary, serving as the outer frame of the machine, while the rotor is movable and positioned within the stator. Both the stator and rotor are typically constructed from ferromagnetic materials. Slots are carved into the inner edge of the stator and the outer periphery of the rotor, where conductors are placed. These conductors are interconnected to form windings, and when voltage is induced in these windings, they are referred to as armature windings. Additionally, there are windings called field windings responsible for generating the main flux. In some machines, permanent magnets may also be utilized to produce the main flux. D.C. machines primarily fall into two categories: D.C. generators and D.C. motors. A D.C. generator converts mechanical energy into electrical energy, while a D.C. motor does the opposite, converting electrical energy into mechanical energy.

D.C. Generator Construction:

A D.C generator mainly consists of three main parts:

1. Magnetic-field system.
2. Armature.
3. Commutator and brushgear.

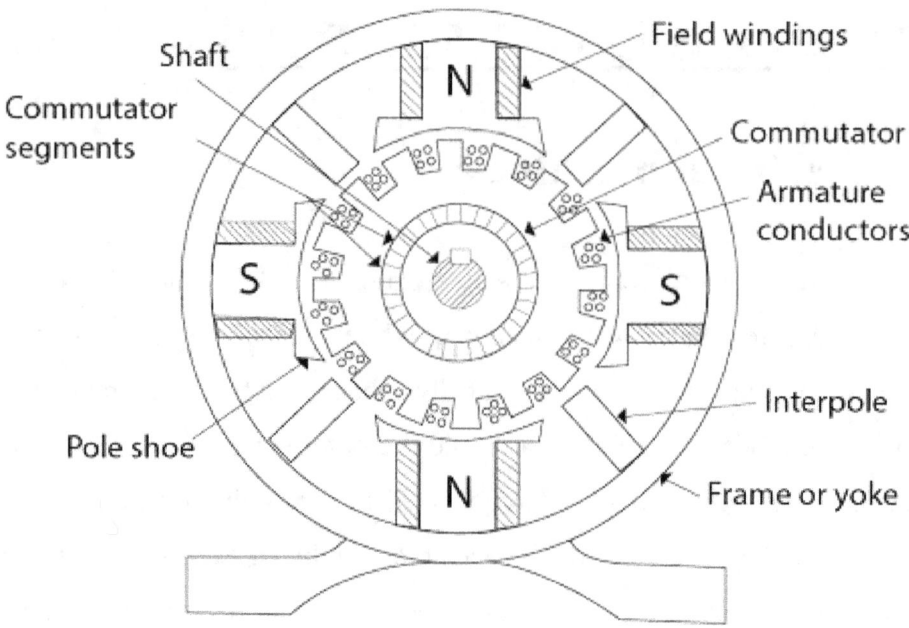

Magnetic-field system:

The stationary part of the machine that produces the main magnetic flux is called the Magnetic-field system. Its outer frame is yoke which is made up of a hollow cylinder of cast steel or rolled steel. The yoke of a D.C. generator is bolted with an even number of pole cores. The yoke serves the following two purposes:

- It acts as a supporter and protector for the pole cores.
- It is used to form the part of a magnetic circuit.

Since the poles point inwards, they are called salient poles. Each pole core has a pole shoe having a curved surface. The pole shoe serves two purposes:

- It supports the field coils.
- It increases the cross-sectional area of the magnetic circuit and reduces its reluctance.

The poles are laminated to reduce the eddy current losses.

Armature:

The armature is the rotating part of the D.C machine. It consists of the shaft upon which a laminated cylinder (armature core) is mounted. The purpose of lamination is to reduce eddy current losses, and the insulated conductor is put in the slot of the armature core, it is called armature winding. The two types of winding used in armature are - **wave** and **lap**.

Commutator and Brush Gear:

The commutator is constructed from numerous wedge-shaped segments crafted from hard drawn copper. These segments are insulated from one another using sheet mica. Shaped like a smooth cylinder, the commutator hosts carbon brushes that gently graze its rounded surface. Both the commutator and the armature are situated on the same shaft. When mechanical power is applied to the machine within a magnetic field, an alternating current electromotive force (e.m.f) is induced in the armature conductor. This e.m.f is then collected by the carbon brushes, indicating that the machine is operating in generator mode. The primary role of the commutator is to convert all negative half cycles into positive ones and to smooth out any ripples to produce a consistent output voltage. Conversely, when the armature is linked to a direct current (d.c.) source via the carbon brushes and commutator, a torque is generated, initiating motor operation. Consequently, the carbon brushes and commutator function as a mechanical rectifier or inverter, facilitating the machine's dual functionality as both a generator and a motor.

The equivalent circuit of a D.C Machine Armature:

The equivalent electric circuit is used to represent the armature of a d.c generator. It can be represented by three series-connected elements E, R_a, and V_b. The element E is the generated voltage, R_a is the armature resistance, and V_b is the brush contact voltage drop. The equivalent circuit of the armature of a d.c generator, d.c motor is shown below. In case of a D.C. motor, E is the back e.m.f.

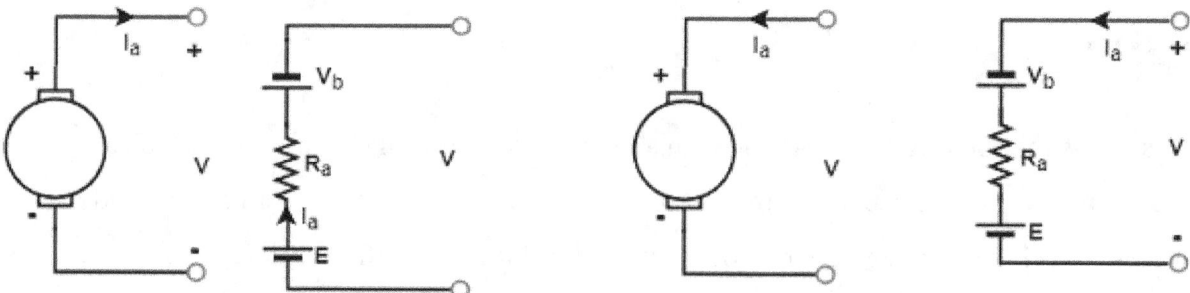

Figure: Equivalent circuits of armature of (a) D.C generator (b) D.C motor

4.1 Speed Control of D.C. Motors

The relationship given below gives the speed of a D.C. motor

$$N = \frac{V - I_a R_a}{k\varphi}$$

The above equation shows that the speed depends upon the supply voltage V, the armature circuit resistance R_a, and the field flux Φ, which is produced by the field current. In practice, the variation of these three factors is used for speed control. Thus, there are three general methods of speed control of D.C. Motors.

1. Resistance variation in the armature circuit: This method is called armature resistance control or Rheostat control.
2. Variation of field flux Φ
   This method is called field flux control.
3. Variation of the applied voltage.
   This method is also called armature voltage control.

1. Armature resistance control (Rheostat Control):

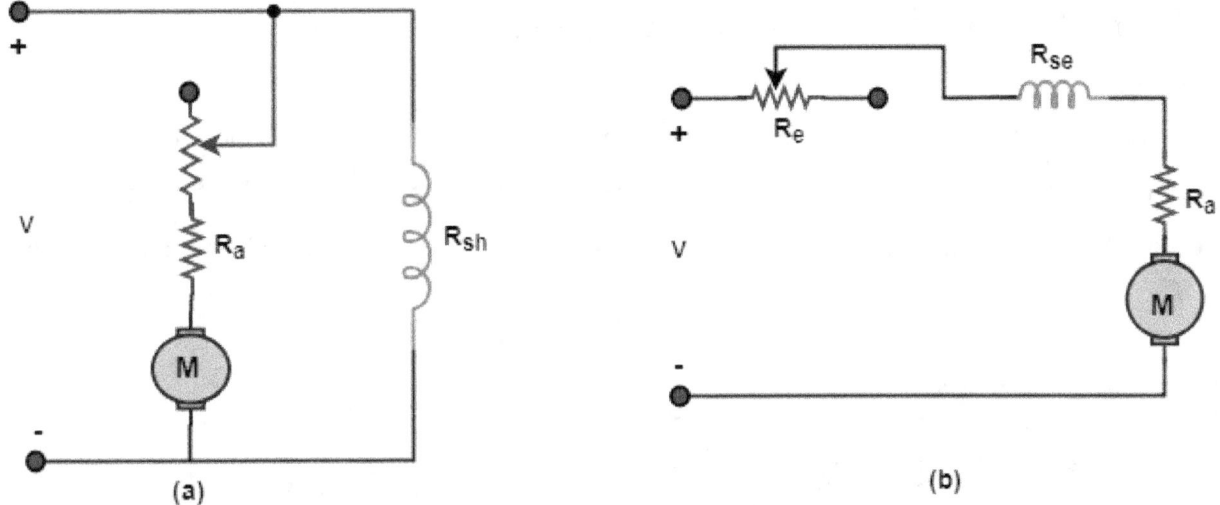

Figure: (a) Speed control of a d.c. Shunt motor by armature resistance control.
(b) Speed control of a D.C. Series motor by armature resistance control.

In this method, a variable series resistor R_e is put in the armature circuit. The figure (a) above shows the process of connection for a shunt motor. In this case, the field is directly connected across the supply and therefore the flux Φ is not affected by variation of R_e.

Figure (b) shows the method of connection of external resistance R_e in the armature circuit of a D.C. series motor. In this case, the current and hence the flux is affected by the variation of the armature circuit resistance.

The voltage drop in R_e reduces the voltage applied to the armature, and therefore the speed is reduced.

This method has the following drawbacks:

1. In the external resistance R_e a large amount of power is wasted.
2. Control is limited to give speed below normal and increase of speed cannot is obtained by this method.
3. For a given value of R_e, the speed reduction is not constant but varies with the motor load.

This method is only used for small motors.

2. Variation of field flux Φ (Field flux control):

Since the field current produces the flux, and if we control the field current then the speed can be controlled. In the shunt motor, speed can be controlled by connecting a variable resistor R_c in series with the shunt field winding. In the diagram below resistor, R_c is called the **shunt field regulator**.

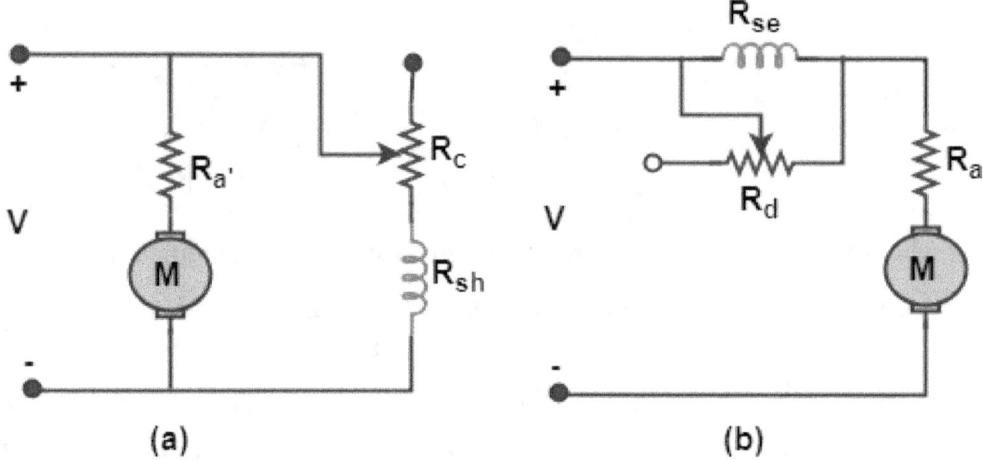

Figure: (a) Speed control of a D.C. shunt motor by variation of field flux.
(b) The diverter in parallel with the series of D.C. Motor.

$$I_{sh} = \frac{V}{R_{sh} + R_c}$$ gives the shunt field current

Any of the one method can vary the field current of the series motor:

- A variable resistance R_d is connected in parallel with the series field winding. The resistor connected in parallel is called the **diverter**. A portion of the main current is diverted through R_d.
- The second method uses a tapped field control.

Here the ampere-turns are varied by varying the number of field turns. This arrangement is used in electric traction.

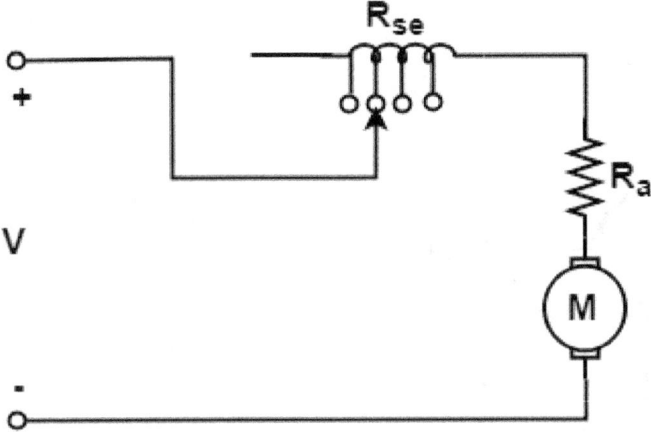

Figure: Tapped series field on D.C. motor

The advantages of field control are as follows:

- This is an easy and convenient method.
- The power loss in the shunt field is small because shunt field current I_{sh} is very small.

3. Armature Voltage control:

We can control the speed of the D.C. motors by varying the applied voltage to the armature. Ward-Leonard system of speed control works on this principle of armature voltage control. In this system, M is the main dc motor whose speed is to be controlled, and G is a separately excited dc generator. The generator G is driven by a 3- phase driving motor which may be an induction motor or asynchronous motor. The combination of ac driving motor and the dc generator is called the motor-generator (M-G) set.

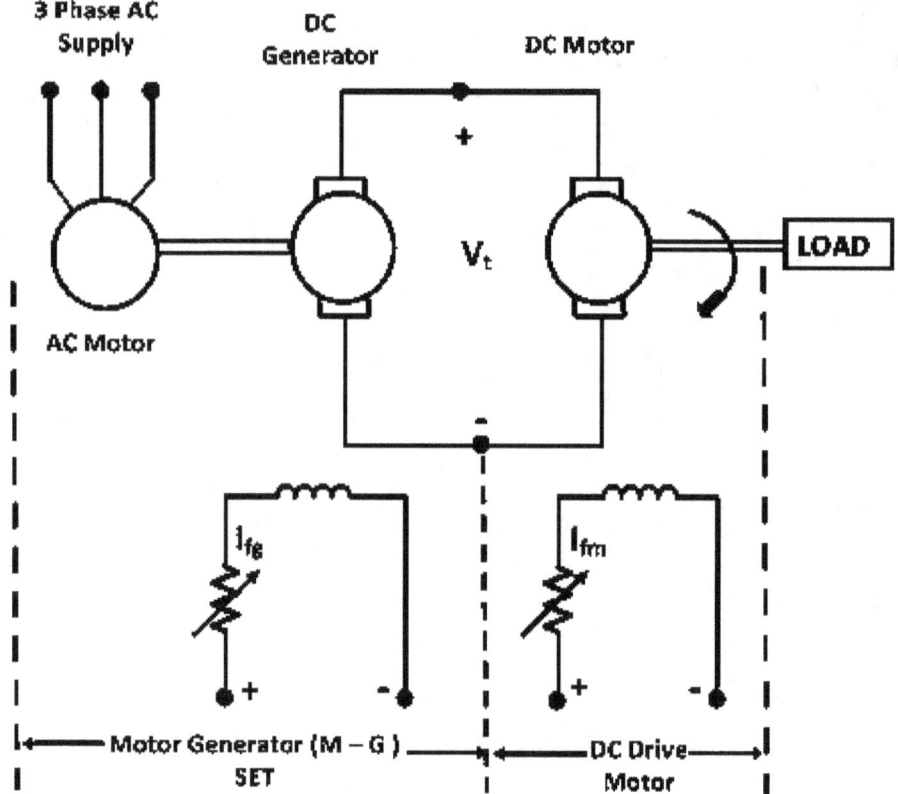

Figure: Ward-Leonard drive

Advantages of Ward-Leonard Drives:

1. This drive has a smooth speed control of dc motors over a wide range in both directions.
2. It has inherent regenerative braking capacity.
3. By using an overexcited synchronous motor as the drive for the dc generator, the lagging reactive volt-amperes of the plant are compensated. Therefore the overall power factor of the plant improves.

Drawbacks of classical Ward-Leonard system:

1. Its initial cost is high because of the use of two additional machines (M-G set) of the same rating as the main dc motor.
2. It has a large size and weight.
3. It requires more floor area and costly foundation.

4. Very frequent maintenance is required.
5. The losses are higher because of lower efficiency.
6. Its drive produces more noise.

4.2 Starting of D.C. Motors

A starter is a device that initiates and accelerates the motor. A controller is a device to start, control speed, reverse, stop and protect the motor.

The motor's armature current is given by

$$I_a = \frac{V - E}{R_a}$$

Thus, the armature current I_a depends upon E and R_a if V is kept constant. When we switch on the motor, the armature is stationary so back e.m.f will be zero. The starting armature current I_{as} is given by

$$I_{as} = \frac{V - 0}{R_a}$$

Three-point starter:

When we start the dc connected motor, the lever turns gradually to the right. When the lever touches the point 1, the field winding gets directly connected across the supply.

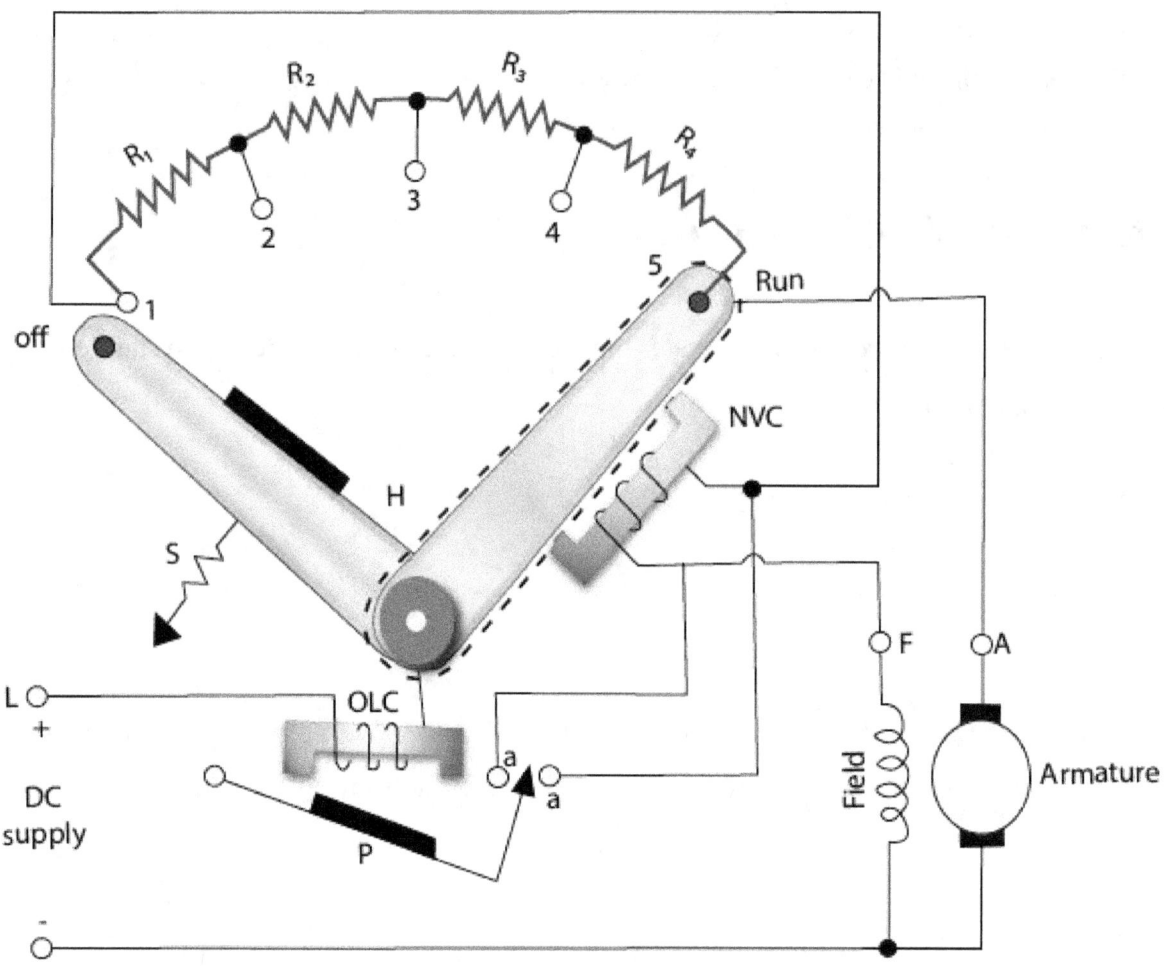

Figure: Three-Point D.C. Shunt motor starter

Working Principle:

A simple method for controlling the current in a machine involves using high-resistance steps placed in series with the armature. In the context of a three-phase shunt motor, a figure illustrates the process of starting the motor. Initially, the handle, labeled as "H," is manually moved, with the first contact point, stud 1, engaged to initiate the starting sequence. In this position, the field winding receives the full supply voltage, while the armature current is regulated through a series of graduated resistances labeled as R1, R2, R3, and R4. The starter handle is then gradually shifted from one step to another until reaching the "RUN" position. At this stage, the motor achieves its full speed, the resistance is completely bypassed, and the supply is directly connected across both windings, ensuring optimal performance.

NVC (No voltage trip coil) is connected in series with field winding of the motor. When the supply voltage falls below a particular value, NVC is, and the handle is pulled back to off position. It also protects against the open circuit of the field winding. The NVC is called no-volt or under voltage protection of the motor.

When armature current exceeds the normal rated value, P is attracted by the electromagnet of OLC (overload coil) and close the contact aa, which short-circuits the NVC. This results in the release of handle H, which return to off position and motor supply is cut off.

To increase the speed of the motor, field resistance should be increased which reduce the current in the shunt field. A very low field current could not hold the switch and handle will achieve the off position. To overcome this difficulty, **4 point starter** is used.

4 point starter:

The main difference between the 3 point starter and 4 point starter is that in 4 point starter the field coil is not connected in series with the no voltage coil that means the holding coil is removed from the shunt field circuit and is connected directly across the line with a current limiting resistance R in series.

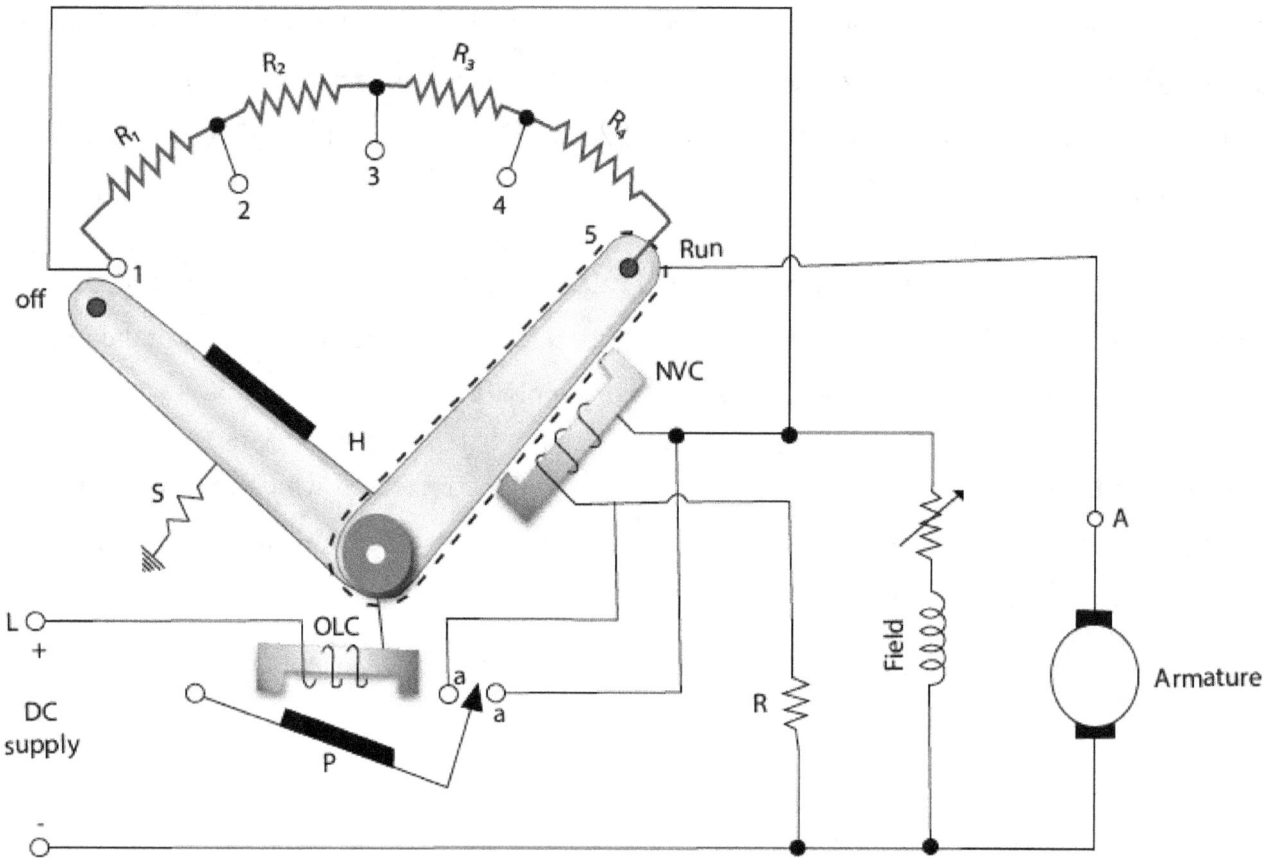

Figure: Four - Point D.C. Shunt motorstarter

In this arrangement three parallel circuits are formed:

1. Armature, overload release and start a resistance.
2. Shunt field winding and variable resistance.
3. Current limiting resistance and holding coil.

With this arrangement, a change in field current for variation of the speed of the motor does not affect the current through the holding coil because the two circuits are independent of each other.

4.3 Types of D.C Machine

The magnetic flux in a d.c machine is produced by field coils carrying current. The production of magnetic flux in the device by circulating current in the field winding is called excitation.

There are two types of excitation in D.C machine. Separate excitation, and self-excitation. In self-excitation, the current flowing through the field winding is supplied by the machine itself, and in separate excitation, the field coils are energized by a separate D.C. Source.

The principal types of D.C machine are:

1. Separately excited d.c. machine
2. Shunt wound or shunt machine.
3. Series wound or series machine.
4. Compound wound or compound machine.

1. Separately excited D.C. machine:

When a separate D.C. source is used to energize the field coils it is called as separately excited D.C. machine. The connections showing the separately excited D.C. Machines are given in the figure.

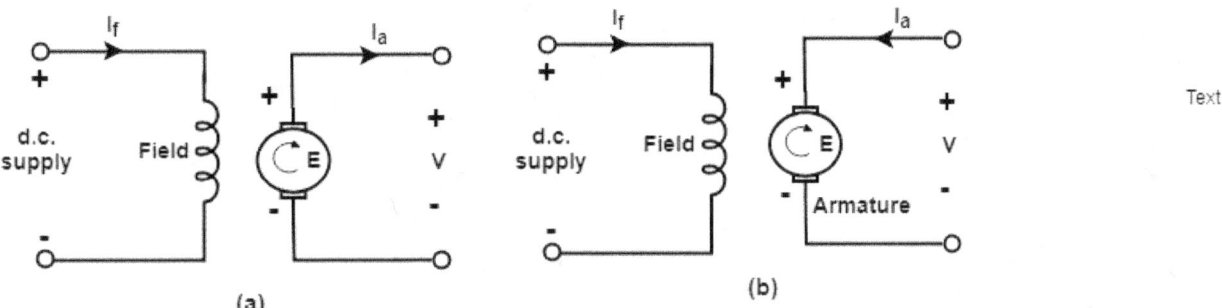

Figure: (a) Separately excited D.C. Generator, (b) Separately excited D.C. Motor.

2. Shunt wound D.C. Machine:

Shunt wound D.C. Machines is the machine in which field coils are connected in parallel with the armature. Since the shunt field receives the full output voltage of a generator or the supply voltage of a motor, it is generally made of a large number of turns of fine wire carrying a small field current.

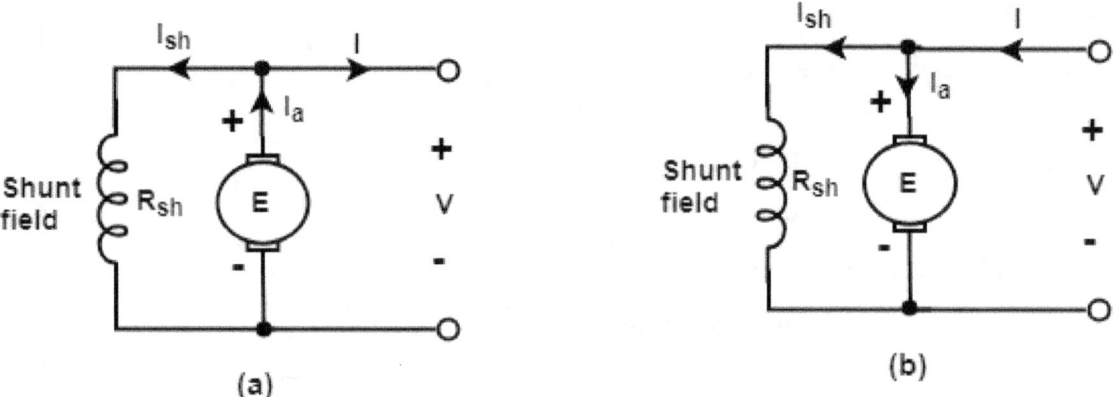

3. Series wound D.C. Machine:

Series wound D.C. Machines is the machine in which the field coils are connected in series with the armature. The series field winding carries the armature current, and the armature current is large, that is why series field winding consists of few turns of wire of large cross-sectional area.

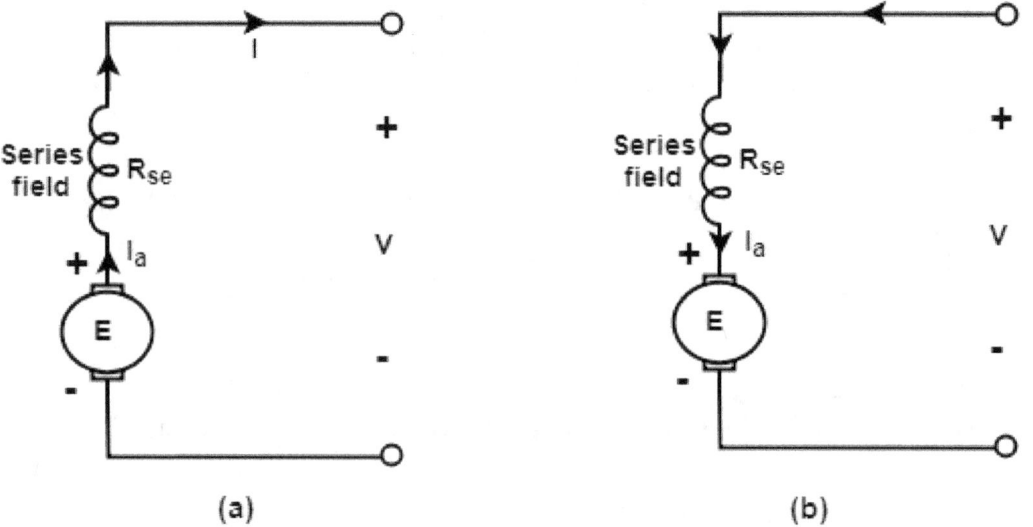

Figure: (a) D.C. series generator (b) D.C. series motor.

4. Compound wound D.C. machine:

A Compound machine is a machine which has both shunt and series fields. Two windings are carried out by each pole of the machine. The series winding has few turns of large cross-sectional area, and the shunt windings have many turns of fine wire.

It can be connected in two ways. If the shunt field is connected in parallel with the armature alone, the machine is called the **short-shunt compound machine** and if the shunt field in parallel with both the armature and series field, the machine is called the **long-shunt compound machine**.

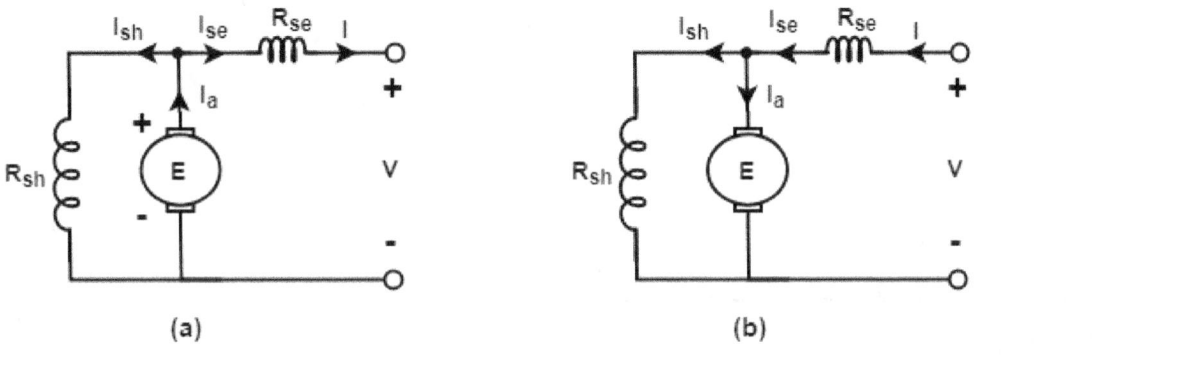

Figure: (a) Short-shunt compound D.C. generator (b) Short-shunt compound D.C. motor.

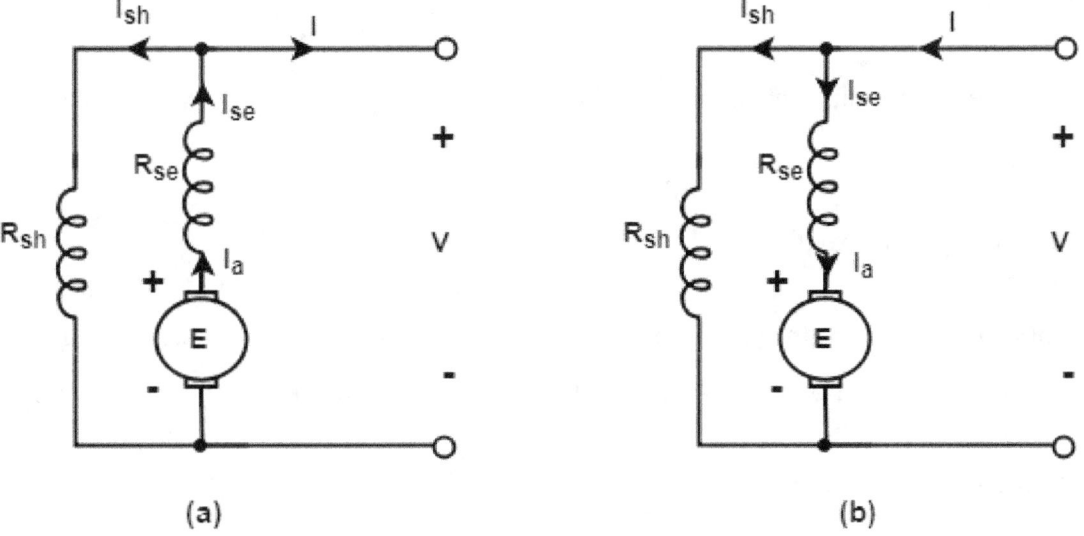

Figure: (a) Long-shunt D.C. generator (b) Long-shunt D.C. motor.

EMF and Torque Equation:

A voltage is generated in the coils when the armature rotates. In case of a generator, the e.m.f. of rotation is called the generated e.m.f and $E_r = E_g$.

In the case of the motor, the e.m.f. of rotation is known as back e.m.f., and $E_r = E_b$. The expression for both the conditions of operation is same.

The clockwise current produces downward field, and anticlockwise current generates the upward field.

Let,

P = Number of poles.

Φ = Flux per pole in Weber.

Z = Total number of the conductor in the armature.

N = Speed of armature

A = No. of parallel paths in the armature.

Let us consider conductor in moving from position P to Q under the pole pitch τP and required to reach from P to Q in t.

Time taken to complete N revolution = 1 min = 60 sec.

Time is taken to complete one revolution = $\dfrac{60}{N}$ sec

During one revolution of the armature in a P pole generator, each armature conductor cuts the magnetic flux P times. So, the flux cut by one conductor in one revolution = ϕP

Average E.M.F induced in one conductor

$$E = \frac{\phi P}{T}$$

$$E = \frac{\phi P}{\left(\frac{60}{N}\right)}$$

$$E = \frac{\phi NP}{60}$$

Total E.M.F generated by Z conductor

$$= \frac{\phi NPZ}{60}$$

E.M.F generated in each parallel path which is connected across a pair of carbon brushes.

E.M.F in each path

$$E = \frac{\phi NPZ}{60A}$$

Torque Equation:

Mechanical power input = ωT ....equation 1

T is the electromagnetic torque developed by the motor running at n r.p.s.

Electrical power developed = $E \times I_a$ ....equation 2

Mechanical power input = electrical power developed

On equating equation 1 and 2

$$\omega T = EI_a$$

Put $E = \frac{\varphi PNZ}{60A}$ and $\omega = 2\pi n = \frac{2\pi N}{60}$

$$\frac{2\pi NT}{60} = \frac{\varphi PNZ}{60A} \cdot I_a$$

$$T = \frac{\varphi PZ}{2\pi A} \cdot I_a$$

The above equation is the torque equation for the D.C. machine.

CHAPTER-5: THREE PHASE INDUCTION MOTORS

Three - Phase induction motor is the most common and popular type of a.c motor used for industrial drives. This motor is used because it is cheap, robust, efficient and reliable. It has good speed regulation and high torque. It requires little maintenance and has a reasonable overload capacity.

Construction

A three-phase induction motor has mainly two parts:

1. Stator.
2. Rotor.

The part of the motor that is fixed means stationary is called the stator.

The rotating part of the motor is called rotor.

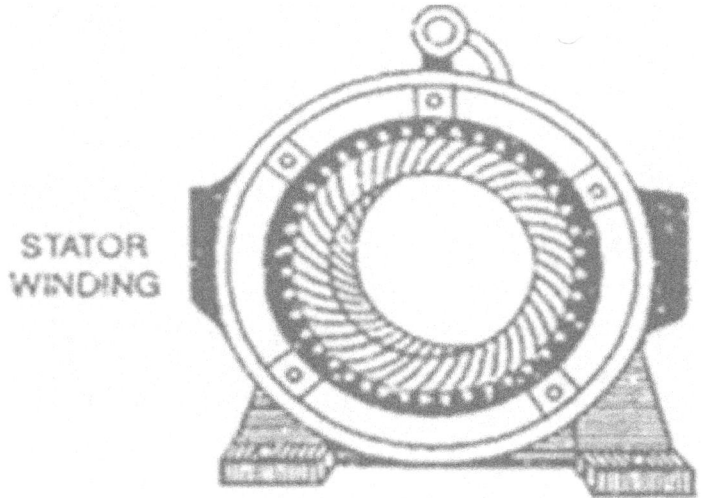

Fig: Induction motor stator with double-layer winding partly wound

The Stator of the motor is built up of high-grade alloy steel laminations, and all the laminations are insulated from each other which are placed on the inner periphery of the system. These laminations are supported on the frame of the cast iron. The conductors of the stator are connected to form three-phase winding either in Star or delta connection.

Types of induction motors:

The induction motors are divided into two parts-

1. Squirrel-cage rotor or simply cage rotor
2. Phase wound or wound rotor. The motors which use this type of rotor are also known as slip-ring motors.

The principle of operation of a Three-phase induction motor

Consider a conductor on the stationary rotor is subjected to a magnetic field that is produced when three phase supply is connected to the three-phase winding of the stator. If the rotation of the magnetic field is clockwise, then it will have the same effect as the conductor moving anticlockwise in the stationary field. By faradays law of electromagnetic induction, a voltage will be induced in the conductor.

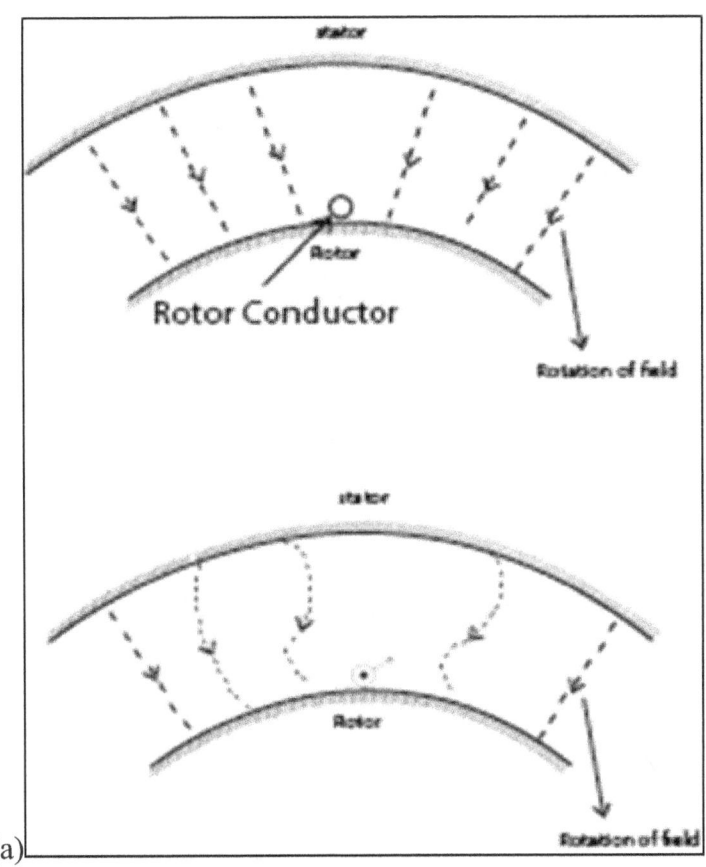

(a)

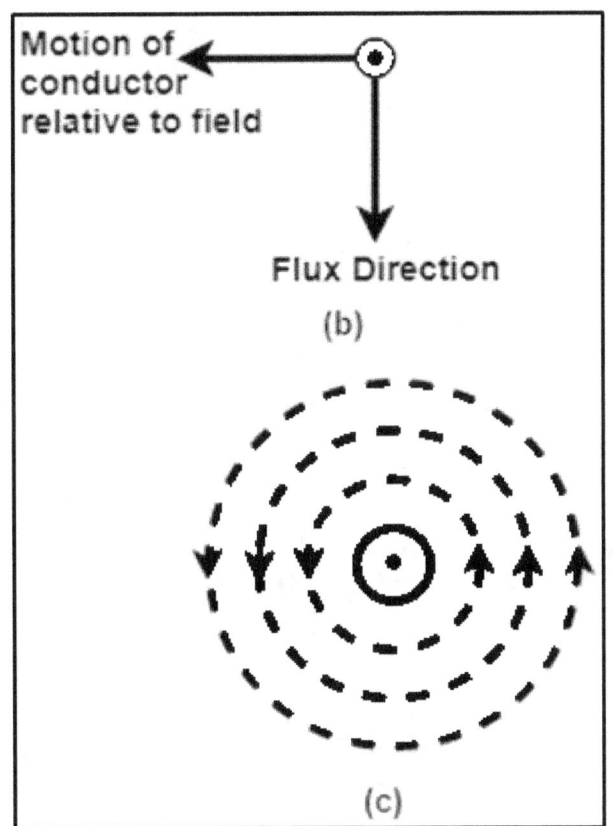

(b)

(c)

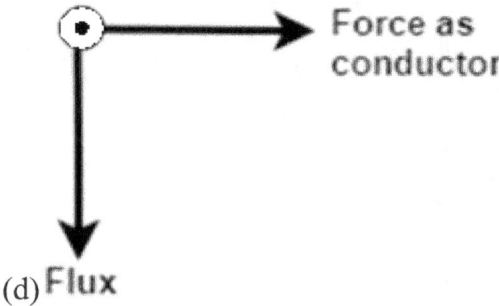

(d) Flux

Fig: Production of Torque

Speed and Slip:

The difference between the synchronous speed and actual speed of the rotor is called the **Slip Speed**.

Therefore we can say that the 'Slip Speed' shows the speed of the rotor relative to the field.

$$S \triangleq \frac{N_s - N_r}{N_s} \text{ Perunit (p.u)}$$

Where,

N_s = synchronous speed in r.p.m

N_r = actual rotor speed in r.p.m

Slip Speed = N_s ?N r.p.m

5.1 Types of three phase induction motor rotor

There are two types of induction motor rotors:

1. Squirrel-cage rotor or simply cage rotor.

2. Phase wound or wound rotors. The motors that use this type of rotor are known as Slip-ring rotors.

Squirrel cage rotor:

The squirrel cage motor operates based on the principles of electromagnetism and comprises essential components such as the rotor, stator, and additional parts like bearings, cylindrical laminated core, and shaft. Bearings play a crucial role in reducing friction between the rotating and stationary parts of the motor. The rotor, made up of a cylindrical laminated core, features parallel slots for accommodating rotor conductors. Unlike wires, these conductors are heavy bars crafted from materials such as copper, aluminum, or alloys. The shaft facilitates the transfer of mechanical power to or from the motor. Meanwhile, the stator serves as the outer stationary component of the motor, completing the essential structure of the squirrel cage motor.

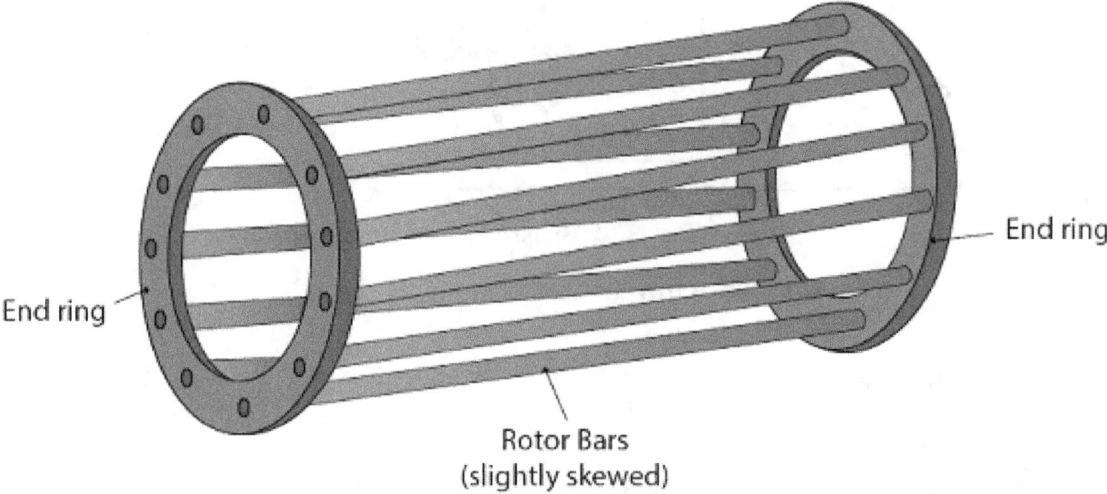

Figure: Cage Rotor

The advantages of skewing of cage rotor conductors are:

1. It helps in reducing noise during the operation and production of uniform torque.
2. During locking, the rotor and stator teeth attract each other due to the magnetic field and this locking tendency is reduced in cage motor.

Wound rotor or slip ring rotor:

The wound rotor is composed of a slotted armature where insulated conductors are inserted into the slots and interconnected to create a three-phase double-layer distributed winding, akin to the stator winding. These rotor windings are arranged in a star configuration. The rotor windings are evenly spread out and typically linked in a star configuration, with their connections brought out of the machine via slip rings situated on the shaft. Copper carbon brushes are utilized to tap into these slip rings. Wound rotor construction is commonly employed for larger machines, particularly when there are demanding starting torque requirements. By incorporating external resistance in the rotor circuit through the slip rings, the starting current can be reduced while also maintaining the starting torque at an optimal level.

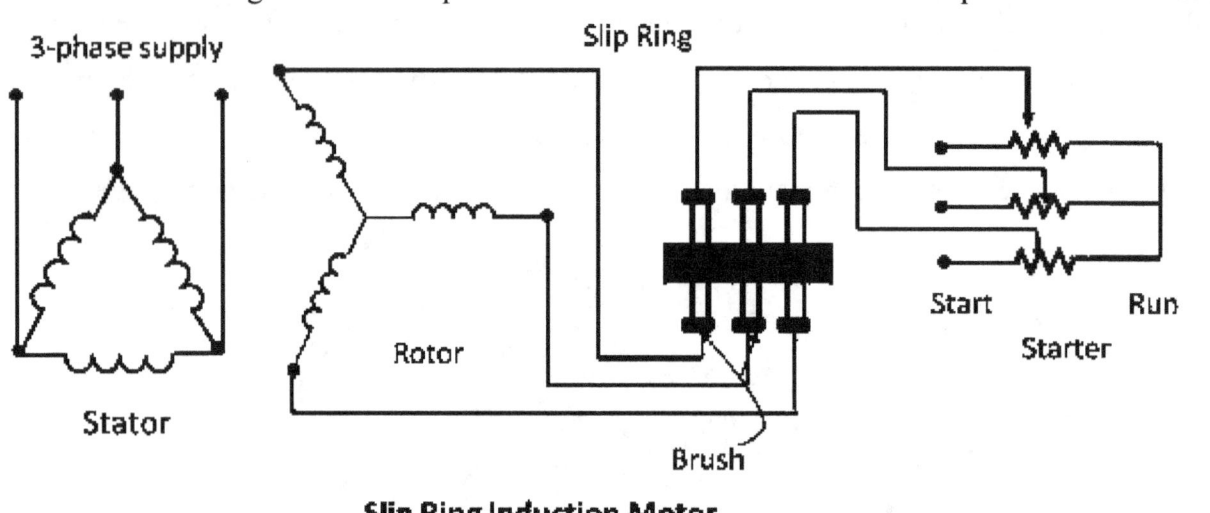

Figure: Slip ring induction motor

Difference between the cage and wound rotors:

The advantages of the cage rotor are as follows:

- Cage rotors have robust construction and are cheaper than wound rotors.
- These rotors do not have brushes because of which the risk of sparking is reduced.
- It requires very less maintenance.
- They have high efficiency and higher power factor.

The advantages of wound rotors are as follows:

- Wound rotors have high starting torque and low starting current in comparison with cage rotors.
- In the case of wound rotors, we can connect additional rotors in the rotor circuit to control the speed.

5.2 No-Load and Blocked Rotor Test

The efficiency of small motors can be determined by directly loading them and by measuring the input and output powers. But in the case of large motors, it is difficult to arrange that much load for them. The power loss will be large if we directly test the load. Therefore indirect methods are used to determine the efficiency of 3-phase induction motors.

We can perform the following test on the motor to find the efficiency:

- No-Load test.
- Blocked-rotor test.

No-Load test or Open-Circuit Test:

The no-load test of an induction motor is same like the open-circuit test of a transformer. In this test, the motor operates without any external load, and the rated voltage at the rated frequency is supplied to the stator. The 2-wattmeter method is commonly used to measure the input power of the system. During this test, a voltmeter measures the standard-rated supply voltage, while an ammeter gauges the no-load current. Because the motor is running without a load, the total power consumed equals the sum of constant losses like iron loss, friction, and winding losses in the motor. $P_{constant} = P_i = P_1 + P_2 = $ Sum of the two wattmeter readings.

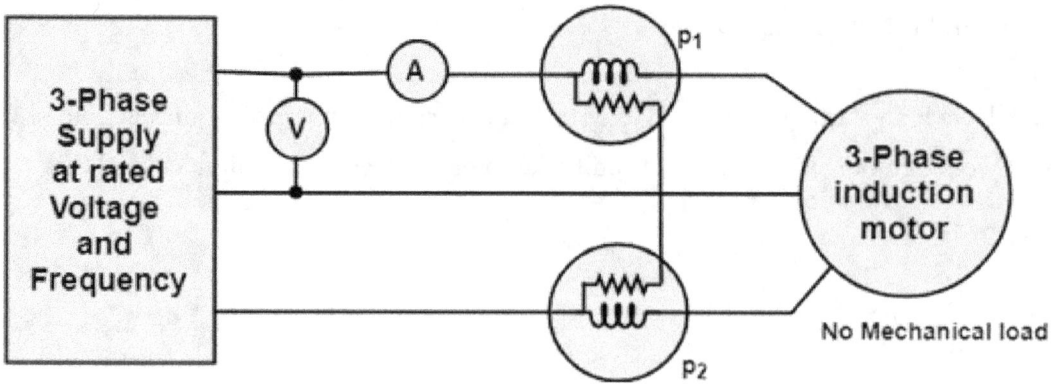

Figure: Circuit diagram for no-load test on a 3-phase induction motor.

Since the power factor of the induction motor under a no-load condition is generally less than 0.5, one wattmeter will show a negative reading. Therefore, it is, necessary to reverse the direction of current-coil terminals to take the reading.

If V_{inl} = input line voltage

P_{inl} = total 3-phase input power at no-load

I_0 = input line current.

V_{ip} = input phase voltage

$P_{inl} = \sqrt{3}\, V_{inl}\, I_0\, \cos\Phi_0$

$I_\mu = I_0 \sin\Phi_0$

$I_\omega = I_0 \cos\Phi_0$

$R_0 = \dfrac{V_{ip}}{I_\omega}$

$X_0 = \dfrac{V_{ip}}{I_\mu}$

Blocked Rotor or Short-Circuit Test:

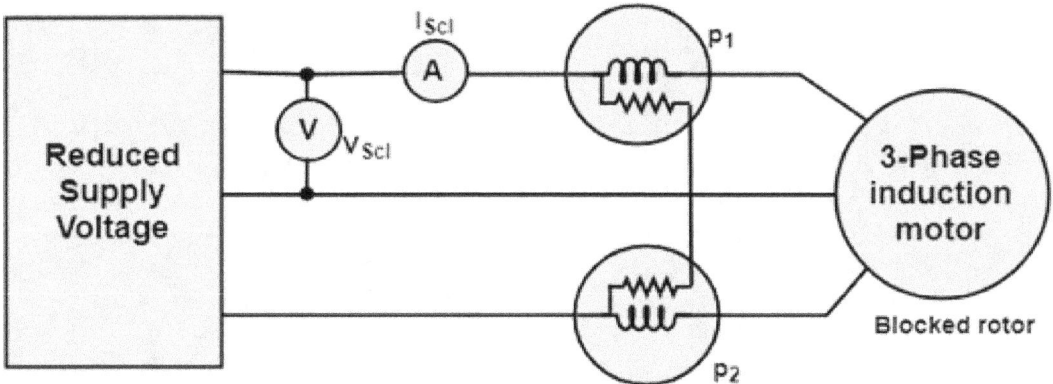

Figure: Circuit diagram for blocked rotor test

The blocked rotor test of an induction motor is similar to the short-circuit test of a transformer. In this test, the shaft of the motor is connected so that it cannot move and rotor winding is short-circuited. In a slip-ring motor, the rotor winding is short-circuited through slip-rings and in cage motors, the rotor bars are permanently short-circuited. This test is also called the locked-Rotor test.

When a reduced voltage at the reduced frequency is applied to the stator through a 3-phase auto-transformer so that full-load current flows in the stator, the following three readings are obtained.

1) The total power input on short-circuits

P_{sc} = algebraic sum of the two wattmeter readings.

2) Reading of ammeter

I_{scl} = line current on short circuit.

3) Reading of voltmeter

V_{scl} = Line voltage on the short circuit

$\therefore P_{sc} = \sqrt{3} V_{scl} \cos?\Phi_{sc}$

Where $\cos? \phi_c$ = Power factor on short circuit

The equivalent resistance of the motor referred to the stator R_{e1} is,

$$R_{e_1} = \frac{P_{scp}}{I_{scp}^2}$$

Equivalent impedance of the motor referred to the stator Z_{e1} is,

$$Z_{e_1} = \frac{V_{scp}}{I_{scp}}$$

Equivalent reactance of the motor referred to stator

$$X_{e_1} = \sqrt{Z_{e_1}^2 - R_{e_1}^2}$$

NOTE: The Blocked-Rotor test should be performed under the same conditions of rotor current and frequency that will exist under normal operating conditions.

5.3 Torque-Slip and Torque-Speed Characteristics

The Torque-slip curve of an induction motor shows the variation of torque with the slip.

We have

$$\tau = \frac{ksR_2 E_{20}^2}{R_2^2 + (sX_{20})^2}$$

If R_2 and X_{20} are kept constant, then the torque τ will depend upon the slip s. The torque-slip characteristic curve is divided into three regions:

- Low-slip region

- Medium-slip region
- High-slip region

(a) Low-slip region

At synchronous speed s = 0, the torque will be zero. When the speed is very near to synchronous speed, the slip is very low and $(sX_{20})^2$ is negligible in comparison with R_2. Therefore,

$$\tau = \frac{k_1 s}{R_2}$$

If R_2 is constant,

$$\tau = \frac{k_3 R_2}{s X_{20}^2}$$

When $k_2 = k_1/R_2$

From the above relation, we can see that the torque is proportional to the slip. Hence, when the slip is small, the torque-slip curve is a straight line.

(b) Medium-slip region

As slip increases, the term $(sX_{20})^2$ becomes large, so that R_2^2 may be neglected in comparison with $(sX_{20})^2$ and

Thus, the torque is inversely proportional to slip towards standstill conditions. We can represent the torque-slip characteristic by a rectangular hyperbola. For intermediate values of the slip, the graph changes from one form to another. In doing so, it passes through the point of maximum torque when $R_2 = sX_{20}$. The maximum torque developed in an induction motor is called the **pull-out torque** or breakdown torque. This developed torque is a measure of the short-time overloading capability of the motor.

(c) High-slip region

The torque decreases beyond the point of maximum torque, and the result is that the motor slows down and then stops. At this stage, we should immediately disconnect the motor from the supply to prevent the damage due to overloading.

The motor operates for the value of the slip between s=0 and s = s_M, where s_M is the value of the slip corresponding to maximum torque. For a typical induction motor, the pull-out torque is 2 to 3 times the rated full-load torque. Thus, the motor can handle short- time overload, without stalling. The starting torque is 1.5 times the rated full-load torque.

The figure drawn below shows the torque-slip curves and torque-speed curves:

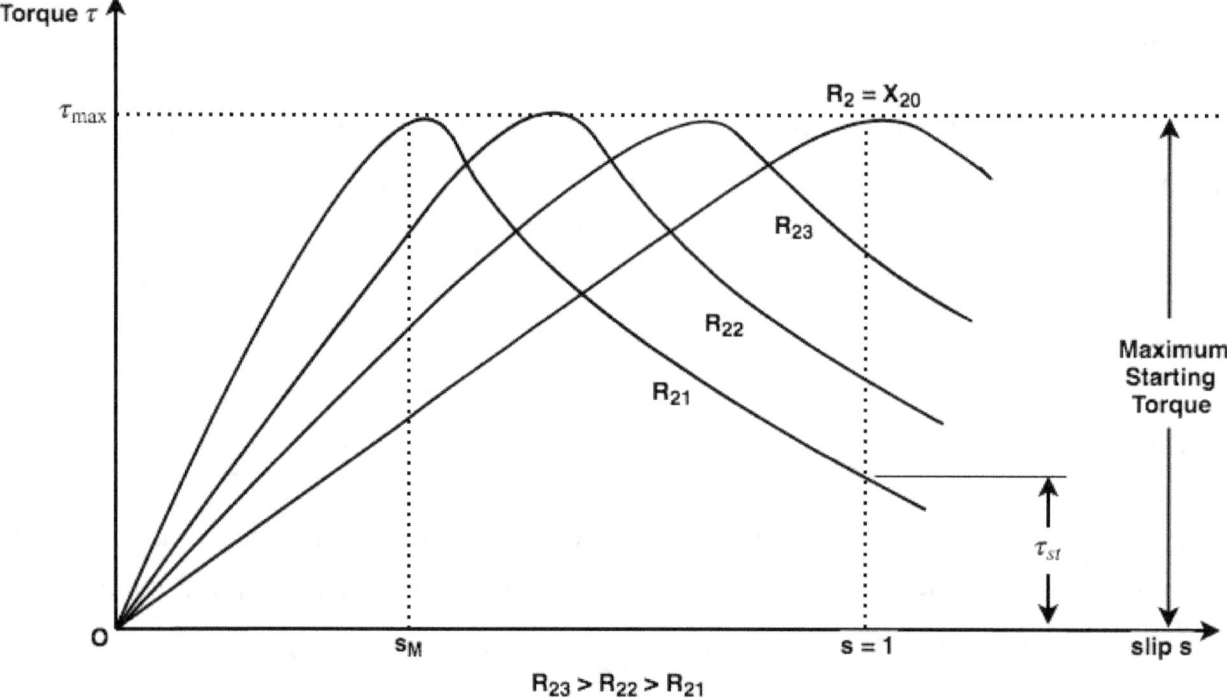

Figure: Torque-slip Curves

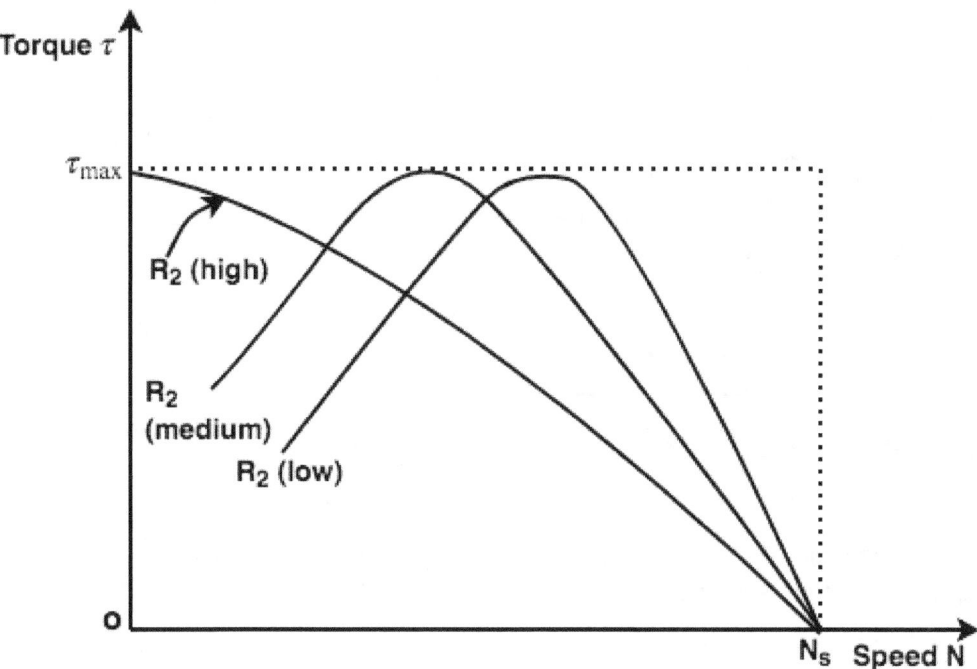

Figure: Torque-speed Curves

5.4 Equivalent circuit of an Induction Motor

Stator's Equivalent Circuit:

The stator model of an induction motor consists of a stator phase winding resistance R_1, a stator phase winding leakage reactance X_1. These two components appear to the right of the machine model. The no-load current I_0 is simulated by a pure inductive reactor X_0 taking the magnetizing component I_μ and a non-inductive resistor R_0 carrying the core-loss current I_ω.

Thus,

$I_0 = I_\mu + I_\omega$

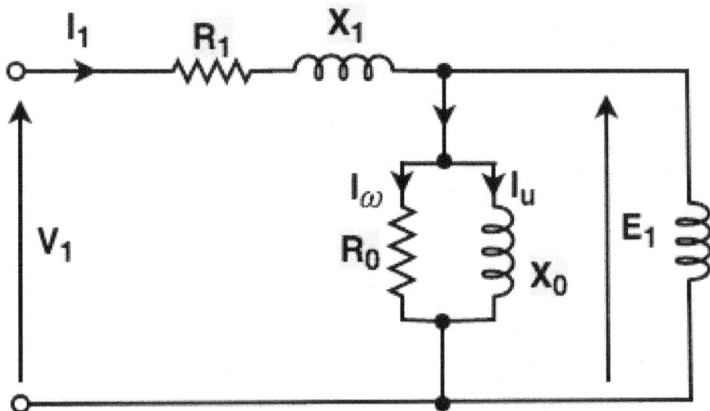

Figure: Stator model of an induction motor

The magnetizing current in the case of the induction motor is higher than the transformer because the air gap of the induction motor causes higher reluctance.

The magnetizing reactance X_0 in an induction motor will have a much smaller value. In a transformer, I_0 is about 2 to 5% of the rated current while in an induction motor it is approximately 25 to 40% of the rated current depending upon the size of the motor.

Rotor's Equivalent Circuit:

It an induction motor, when a 3φ supply is applied to the stator windings, a voltage is induced in the rotor windings of the machine. In general, the higher the relative motion of the rotor and the stator magnetic fields, the higher is the resulting rotor voltage. The maximum relative motion occurs when the rotor is stationary, this condition is called the Standstill condition. This is also known as the locked-rotor or blocked-rotor condition.

If the induced rotor voltage is E_{20} then the induced voltage at any slip is given by,

$E_{2s} = {}_sE_{20}$

The rotor resistance R_2 is constant. It is independent of slip.

The reactance of the induction motor rotor depends upon the inductance of the rotor, voltage frequency, and rotor's current.

If L_2 = inductance of rotor, the rotor reactance is given by

$$X_2 = 2\pi f_2 L_2$$

But $f_2 = sf_1$

∴ $X_2 = 2\pi \, sf_1 \, L_2 = s(2\pi f_1 L_2)$

Or $X_2 = sX_{20}$

In the above equation, X_{20} is the standstill reactance of the rotor.

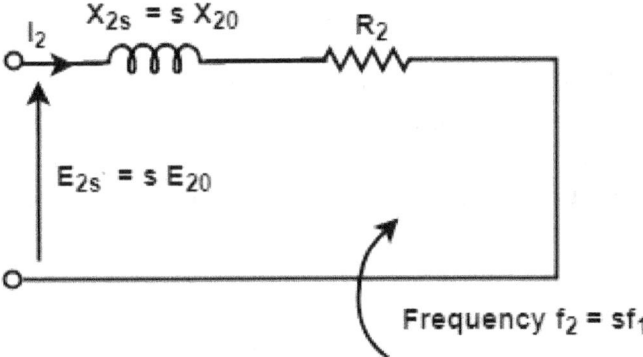

Figure: Rotor Circuit Model

The rotor impedance is given by

$Z_{2s} = R_2 + jX_{2s}$
Or $Z_{2s} = R_2 + jsX_{20}$

The rotor current per phase is given as

$$I_{2s} = \frac{E_{2s}}{Z_{2s}}$$

$$I_{2s} = \frac{sE_{20}}{R_2 + jsX_{20}}$$

In the circuit drawn above, I_2 is a slip-frequency current produced by a slip-frequency induced voltage sE_{20} acting in the rotor circuit having an impedance per phase of $(R_2 + jsX_{20})$.

By dividing both the numerator and the denominator of the above equation by the slip s, we get

$$I_{2s} = \frac{E_{20}}{\frac{R_2}{s} + jX_{20}}$$

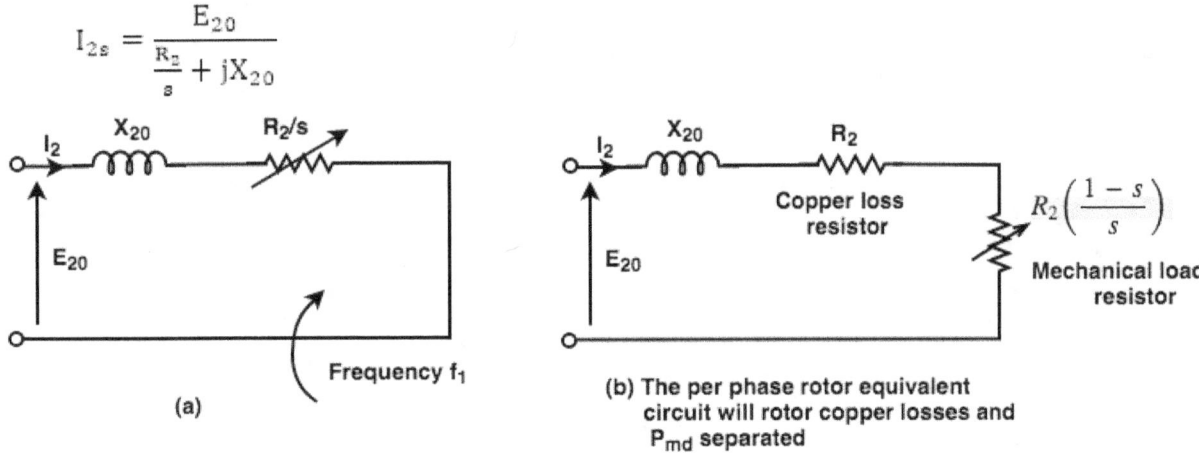

(a) Frequency f_1

(b) The per phase rotor equivalent circuit will rotor copper losses and P_{md} separated

It is to be noted that the magnitude and phase angle of I2s remain the same by this operation.

This equation describes the secondary circuit of a fictitious transformer, one with a constant voltage ratio and with the same frequency of both sides. This fictitious stationary rotor carries the same current as the actual rotating rotor, and, thus produces the same m.m.f wave. This concept of fictitious stationary rotor makes it possible to transfer the secondary impedance to the primary side.

CHAPTER-6: PARALLEL OPERATIONS OF TRANSFORMERS

When the primary windings of two transformers are connected to a common supply voltage and the secondary windings of both the transformers to a common load, this type of connection of transformer is said to be the **parallel operation of transformers**.

Reasons for parallel operation

The reasons for operating the transformers in parallel are as follows:

1. This is an economical method because a single large transformer is uneconomical for large load.
2. If the transformers are connected in parallel, we require extra load then we can expand the system by adding more transformers in the future.
3. Parallel operation reduces the space capacity of the substation when we connect transformers of standard size.
4. The parallel connection maximizes the electrical power system availability as we can shut down any system for maintenance without affecting other systems performance.

Single-phase transformers in parallel:

The diagram drawn below shows the circuit diagram of two transformers A and B connected in parallel.

Let,

a_1 = turns ratio of transformer A

a_2 = turns ratio of transformer B

Z_A = equivalent impedance of transformer A referred to the secondary side.

Z_B = equivalent impedance of transformer B referred to the secondary side.

Z_L = load impedance across the secondary side.

I_A = current supplied to the load by the secondary of transformer A.

I_B = current supplied to the load by the secondary of transformer B.

V_L = load secondary voltage.

I_L = load current

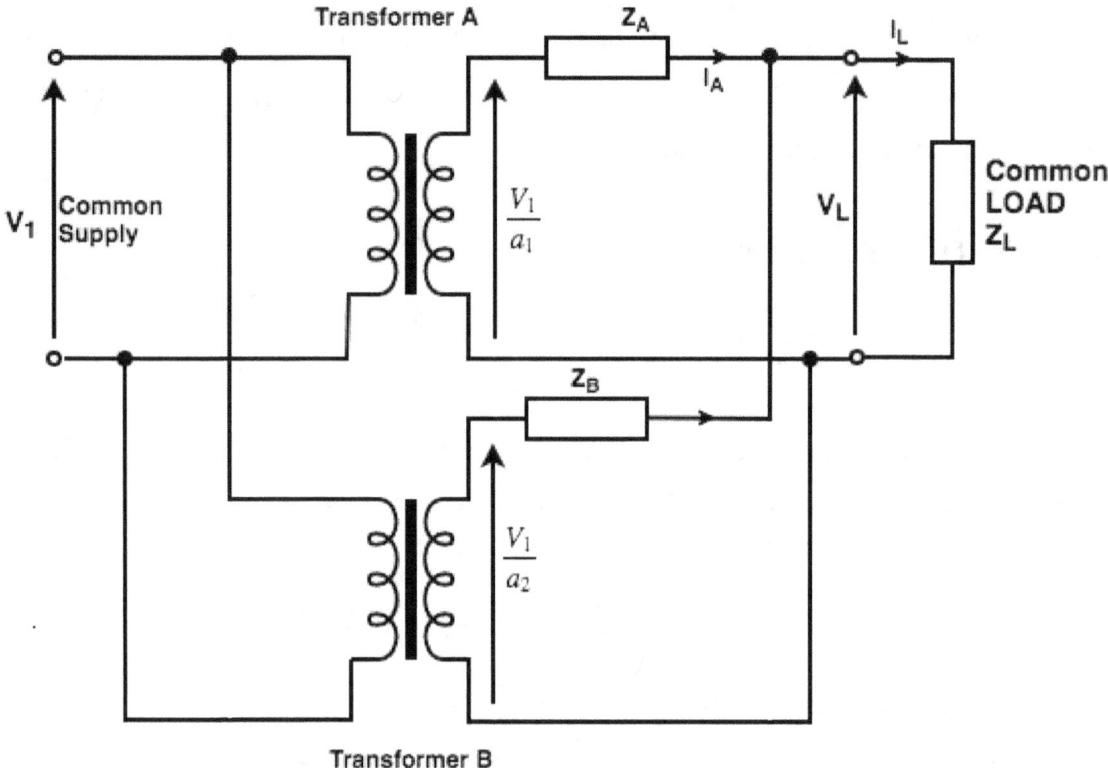

Fig: Two single-phase transformers in parallel.

By KCL,

$I_A + I_B = I_L$

BY KVL,

$$V_L = \frac{V_1}{a_1} - I_A Z_A$$

$$V_L = \frac{V_1}{a_2} - I_B Z_B = \frac{V_1}{a_2} - (I_L - I_A) Z_B$$

By solving the above two equations, we get

$$I_A = \frac{Z_B I_L}{Z_A + Z_B} + \frac{V_1(a_2 - a_1)}{a_1 a_2 (Z_A + Z_B)}$$

$$I_B = \frac{Z_A I_L}{Z_A + Z_B} - \frac{V_1(a_2 - a_1)}{a_1 a_2 (Z_A + Z_B)}$$

Each of these currents has two components; the first component represents the transformer's share of the load current and the second component is a circulating current in the secondary windings.

Circulating currents have the following undesirable effects:

- They increase the copper loss.
- They overload one transformer and reduce the permissible load KVA.

Conditions for parallel operation of Single-Phase transformers:

Necessary conditions

1. The transformers must have the same polarities.
2. The transformers should have equal turn ratios.

Desirable conditions

1. The voltages at full load across transformers internal impedance should be equal.
2. The ratios of their winding resistances to reactances should be equal for both transformers. This condition ensures that both transformers operate at the same power factor, thus sharing active power and reactive voltamperes according to their ratings.

Three-phase transformers in parallel

The conditions for proper parallel operation of single-phase transformers are as follows:

1. The polarities of the transformers should be the same.

2. Identical primary and secondary voltage ratings.
3. Impedances inversely proportional to the kVA ratings.
4. Identical X/R ratios in the transformer impedances.

The condition for the parallel operation of single phase and three phase transformer is the same but with the following additions:

1. The phase sequence of the transformers must be identical.
2. The primary and secondary voltages of all the transformers connected in parallel must have the same phase shift.

NOTE: Under-balanced loading conditions, the three-phase transformer calculations are made on a per-phase basis. It is, however, preferable to perform calculations on a per-unit basis, particularly in cases where the primary and secondary connections are different.

CHAPTER-7: REGULATION AND EFFICIENCY

What is voltage Regulation?

Voltage regulation refers to the property that describes how the voltage of a transformer changes as it is loaded. The voltage regulation of the transformer is defined as the arithmetical difference in the secondary terminal voltage between no-load ($I_2=0$) and full rated load ($I_2 = I_{2fl}$) at a given power factor with the same value of primary voltage for both rated load and no-load.

The numerical difference between no-load and full-load voltage is called **inherent voltage regulation**.

Inherent voltage regulation

$$\triangleq |V_{2nl}| - |V_{2fl}|$$

Where V_{2fl} = rated secondary terminal voltage at rated load.

V_{2nl} = no load secondary terminal voltage with the same value of primary voltage for both rated load and no load.

Per unit voltage regulation at full load is

$$\triangleq \left. \left| \frac{|V_{2nl}| - |V_{2fl}|}{|V_{2fl}|} \right| \right|_{|V_1| = \text{constant}}$$

Percent voltage regulation at full load

$$\triangleq \left. \left| \frac{|V_{2nl}| - |V_{2fl}|}{|V_{2fl}|} \times 100 \right| \right|_{|V_1| = \text{constant}}$$

We can say that voltage regulation is an important measure for the performance of the transformer. We can specify the limits of the transformer in terms of voltage regulation.

Voltage Regulation in terms of primary values:

Per unit voltage regulation

$$= \frac{||V_{2nl}| - |V_{2fl}||}{|V_{2fl}|}$$

$$V_{2nl} = \frac{V_1}{a}$$

∴pu voltage regulation is

$$= \frac{\left|\frac{V_1}{a}\right| - |V_{2fl}|}{|V_{2fl}|}$$

Calculation of Voltage Regulation

We can calculate the voltage regulation of a transformer in terms of circuit parameters. The equivalent circuit diagram of a transformer referred to secondary is shown below.

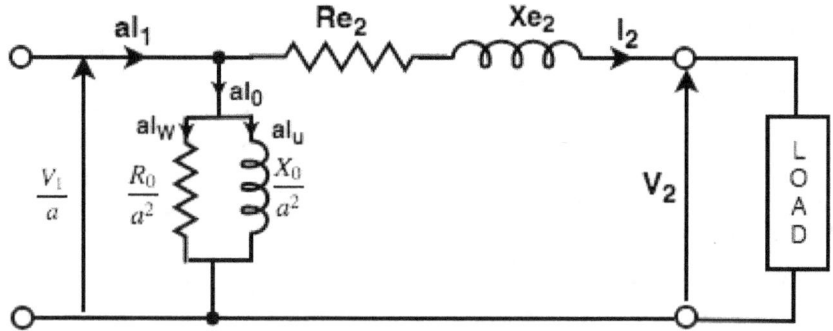

Figure: Equivalent circuit of the transformer referred to the secondary

By KVL, $$\frac{V_1}{a} = V_2 + I_2 Z_{e_2}$$

To calculate the voltage regulation we have to follow the steps given below:

1. We have to take V_2 as the reference Phasor

$\therefore V_2 = V_2 \angle 0° = V_2 + j0$

2. Write I_2 in Phasor form

For lagging power-factor $\cos\Phi_2$

$I_2 = I_2 \angle -\Phi_2 = I_2 \cos\Phi_2 - jI_2 \sin\Phi_2$

For leading power-factor $\cos\Phi_2$

$I_2 = I_2 \angle +\Phi_2 = I_2 \cos\Phi_2 + jI_2 \sin\Phi_2$

For unity power factor

$I_2 = I_2 \angle 0° = I_2 + j0$

3. Calculate Z_{e2})

$Z_{e2} = R_{e2} + jX_{e2}$

4. Calculate

$$V_{2nl} = \frac{V_1}{a}$$

$$\frac{V_1}{a} = V_2 \angle 0° + I_2 Z_{e_2}$$

5. Now we will calculate the voltage regulation

$$= \frac{\left|\frac{V_1}{a}\right| - |V_{2fl}|}{|V_{2fl}|} \text{ pu}$$

Or we can say that it is the ratio of the difference between the voltages at no load and full load to the voltage at full load.

What is Transformer Efficiency?

Transformer efficiency (η) can be explained as the ratio of the output power to the input power.

$$\eta \triangleq \frac{\text{output power}}{\text{input power}}$$

$$= \frac{\text{output power}}{\text{output power + copper loss + iron loss}} \text{ pu}$$

Therefore the per unit efficiency at load current I_2 and power factor $\cos \Phi_2$ will be

$$\eta = \frac{V_2 I_2 \cos\phi_2}{V_2 I_2 \cos\phi_2 + I_2^2 \cdot R_{e_2} + P_i} \text{ pu}$$

And the per unit efficiency at full load is

$$\eta_{fl} = \frac{V_2 I_{2fl} \cos\phi_2}{V_2 I_{2fl} \cos\phi_2 + I_{2fl}^2 \cdot R_{e_2} + P_i}$$

If $S_{2fl} = (VA)_{2fl} = V_2 I_{2fl}$ = full-load VA = rated VA

Then

$$\eta_{fl} = \frac{S_2 \cos\phi_2}{S_2 \cos\phi_2 + P_{cfl} R_{e_2} + P_i}$$

We know that in a transformer there is no rotational part so there are no rotational losses such as windings and frictional losses in a rotating machine. Therefore, we can obtain a maximum efficiency as high as 99% in a well-designed transformer.

What is all day (or energy) efficiency of a transformer?

The ratio of the energy output of 24-hour to the energy input of a 24-hour period is called the All-day efficiency of a transformer.

$$\eta_{AD} = \frac{\text{energy output over 24 hours}}{\text{energy input over 24 hours}}$$

$$= \frac{\text{energy output over 24 hours}}{\text{energy output over 24 hours} + \text{energy losses over 24 hours}}$$

If we know the load cycle of the transformer, then all day efficiency can easily be determined.

CHAPTER-8: TYPES OF SINGLE PHASE INDUCTION MOTORS

The single-phase induction motor is started by using some methods. Mechanical methods are not very practical methods that is why the motor is started temporarily by converting it into a two-phase motor.

Single-phase induction motors are classified according to the auxiliary means used to start the motor. They are classified as follows:

1. Split-phase motor
2. Capacitor-start motor
3. Capacitor-start capacitor-run motor
4. Permanent-split capacitor (PSC) motor
5. Shaded-pole motor

1. Split-phase induction motor:

The split-phase induction motor is also known as a **resistance-start motor**. It consists of a single-cage rotor, and its stator has two windings ? the main winding and a starting (also known as an auxiliary) winding. Both the windings are displaced by 90° in space like the windings in a two-phase induction motor. The main winding of the induction motor has very low resistance and high inductive reactance.

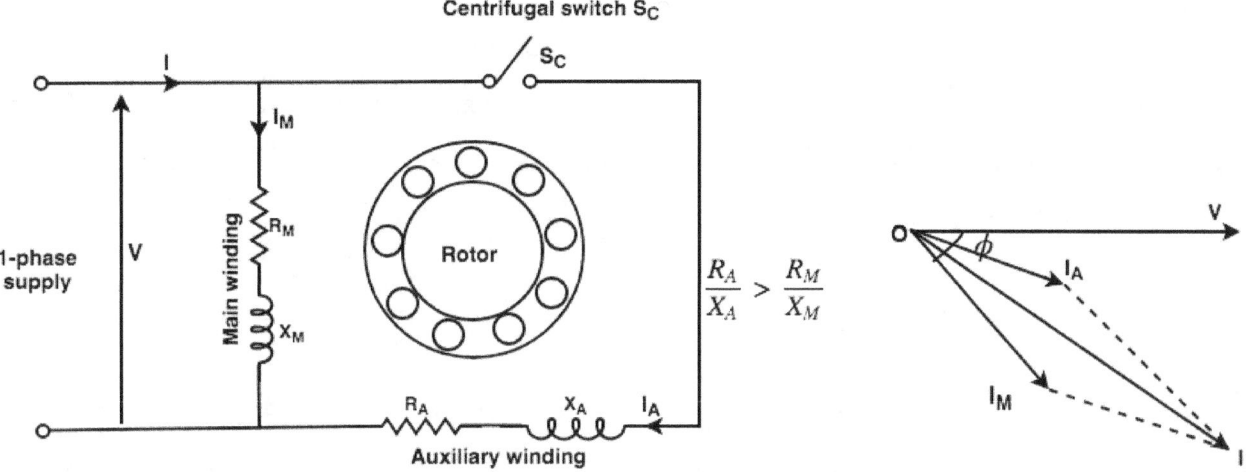

Figure: Split-phase induction motor (a) Circuit diagram (b) Phasor diagram

Motor Characteristics:

The starting torque of a resistance-start induction motor is about 1.5 times full-load torque. The maximum or pull-out torque is about 2.5 times full-load torque at about 75% of synchronous speed. The split-phase motor has a high starting current which is usually 7 to 8 times the full-load value.

Applications:

Split-phase motors are most suitable for easily started loads where the frequency of starting is limited, and these are very cheap.

1. These motors are used in washing machines.
2. These are used in Air conditioning fans.
3. Used in food mixers, grinders, floor polishers, blowers, centrifugal pumps,
4. These are used in small drills, lathes, office machinery, etc.
5. Sometimes they are also used for drives requiring more than 1kW.

Capacitor motors:

Capacitor motors are the motors that have a capacitor in the auxiliary winding circuit to produce a greater phase difference between the current in the main and auxiliary windings. There are three types of capacitor motors.

2. Capacitor-start motor:

The capacitor-start motor develops a much higher starting torque, i.e. 3.0 to 4.5 times the full-load torque. To obtain a high starting torque, the value of the starting capacitor must be large, and the resistance of starting winding must be low. Because of the high VAr rating of the capacitor required, electrolytic capacitors of the order of 250 µF are used. The capacitor Cs is short-time rated.

These motors are more costly than split-phase motors because of the additional cost of the capacitor.

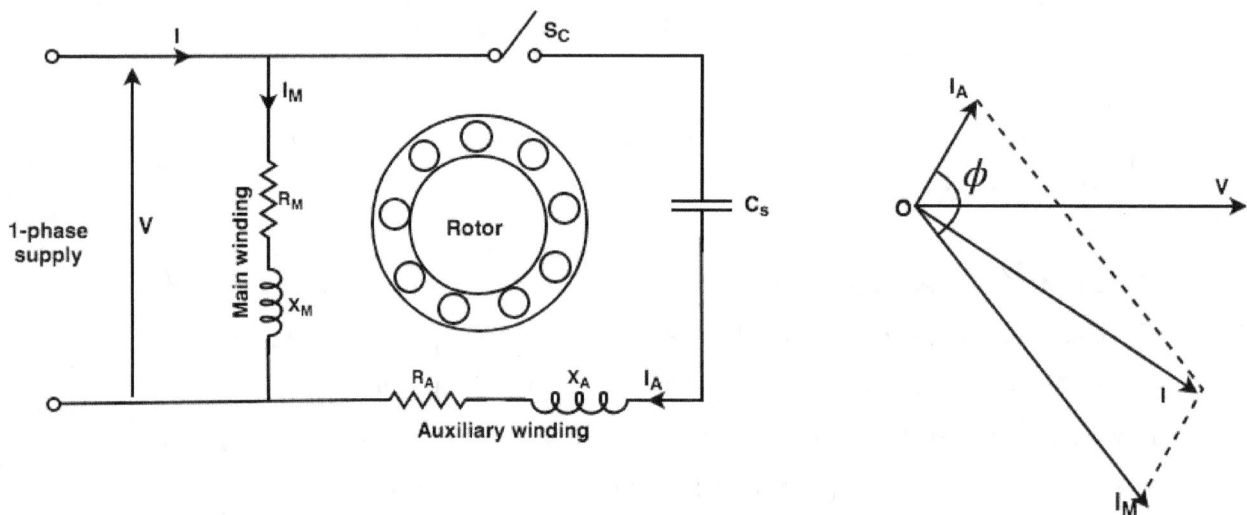

Figure: Capacitor start motor (a) circuit diagram (b) Phasor diagram

Applications:

1. These motors are used for heavy loads where frequent start required.
2. These motors are used for pumps and compressors, so these are used as a compressor in the refrigerator and air conditioner.
3. They are also used for conveyors and some machine tools.

3. Two-Value Capacitor Motor

This motor has a cage rotor, and its stator has two windings namely the main winding and the auxiliary winding. The two windings are displaced 90?in space. The motor uses two capacitors Cs and CR. In the initial stage, the two capacitors are connected in parallel.

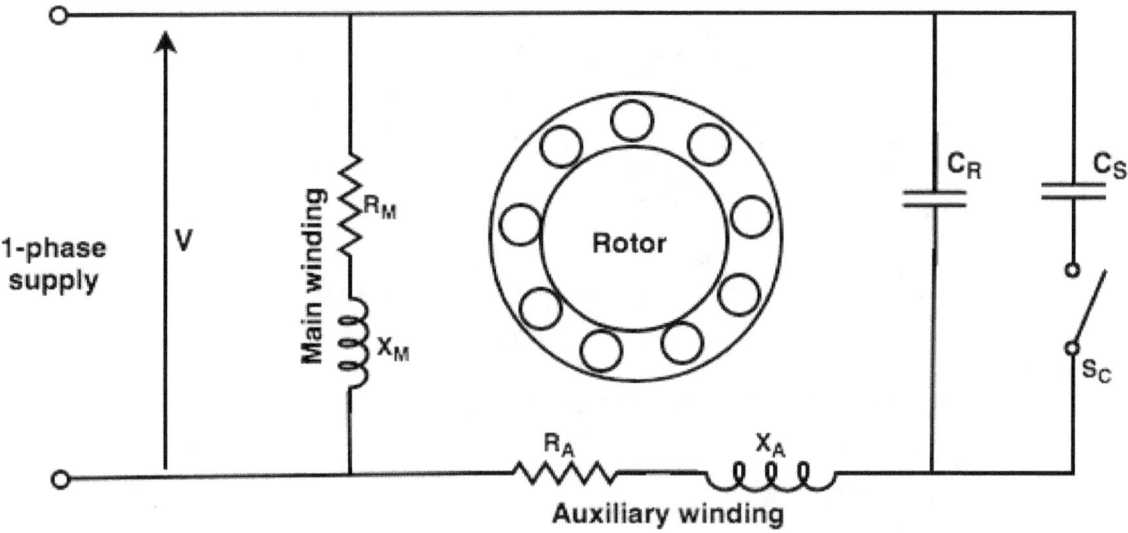

Figure: Two-value capacitor motor

Applications:

1. Two value capacitor motors are used for loads of higher inertia that requires frequent start.
2. These are used in pumping equipment.
3. These are used in refrigeration, air compressors, etc.

4. Permanent-split Capacitor (PSC) motor:

These motors have a cage rotor, and its rotor consists of two windings namely, the main winding and the auxiliary winding. The single-phase induction motor has only one capacitor C which is connected in series with the starting winding. The capacitor C is permanently connected in series with the starting winding. The capacitor C is permanently connected in the circuit at starting and running conditions.

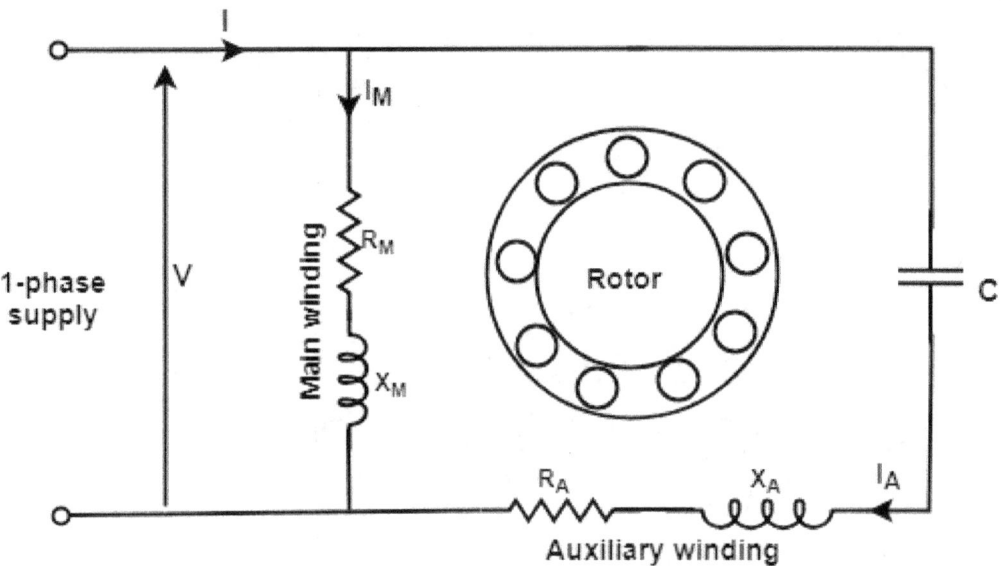

Figure: Permanent-split capacitor motor

Advantages

A single-value capacitor motor has the following advantages:

1. In this type of motor, no centrifugal switch is required.
2. This motor has higher efficiency.
3. It has higher power-factor because of a permanently-connected capacitor.
4. It has higher pull-out torque.

Limitations of permanent-split capacitor motor:

1. Electrolytic capacitors cannot be used for continuous running. Therefore, paper-spaced oil-filled type capacitors are to be used. Paper capacitors of the same rating are larger in size and more costly.
2. A single-value capacitor has a low starting torque usually less than full-load torque.

Applications:

1. These motors are used for fans and blowers in heaters.
2. It is used in air conditioners.
3. It is used to drive refrigerator compressors.
4. It is also used to operate office machinery.

5. Shaded pole motor:

A shaded-pole motor is a simple type of self-starting single-phase induction motor. It consists of a stator and a cage-type rotor. The stator is made up of salient poles. Each pole is slotted on the side, and a copper ring is fitted on the smaller part. This part is called the shaded pole. The ring is usually a single-turn coil and is known as shading coil.

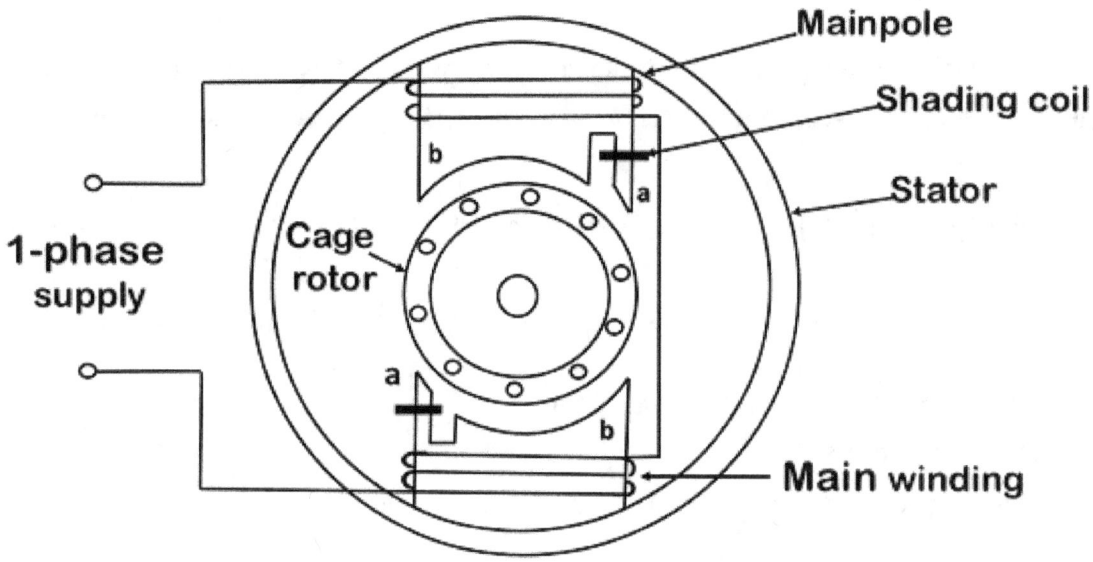

Figure: Shaded-pole motor with two stator poles.

Applications:

1. Shaded-pole motors are used to drive devices which require low starting torque.
2. These motors are very suitable for small devices like relays, fans of all kinds, etc. because of their low initial cost and easy starting.
3. The most common application of these motors is in table fans, exhaust fans, hair dryers, fans for refrigeration and air-conditioning equipment, electronic equipment, cooling fans, etc.

8.1 Working Principle of a Single Phase Induction Motor

Production of Rotating Field

Consider two winding 'A' and 'B' so displaced that they produce magnetic field 90° apart in space. The resultant of these two fields is a rotating magnetic field of constant magnitude ϕ_m.

Non-Uniform magnetic field produces a non-uniform torque which makes the operation of the motor noisy, affect starting torque.

$\varphi_A = \varphi_m \sin \omega t$

$\varphi_B = \varphi_m \sin(\omega t + 90°)$

Figure: Production of the uniform magnetic field.

Starting Principle

A single phase induction motor consists of a single phase winding on the stator and a cage winding on the rotor. When a 1 phase supply is connected to the stator winding, a pulsating magnetic field is produced. In the pulsating field, the rotor does not rotate due to inertia. Therefore a single phase induction motor is not self-starting and requires some particular starting means. Two theories have been suggested to find the performance of a single phase induction motor.

1. Double revolving field theory.
2. Cross-field theory.

Double revolving field theory

This theory for single phase states that a stationary pulsating magnetic field can be resolved into two RMF, each of equal magnitude but rotating in the opposite direction.

The induction machine responds to each magnetic field separately, and the net torque in the motor is equal to some of the torque due to each of the two magnetic fields.

The equation for an alternating magnetic field whose axis is fixed in space is given by:

$$b(\alpha) = \beta_{max} \sin \omega t \cos \alpha$$

$$b(\alpha) = \frac{1}{2}\beta_{max} \sin(\omega t - \alpha) + \frac{1}{2}\beta_{max} \sin(\omega t + \alpha)$$

β_{max} is the maximum value of sinusoidally distributed air gap flux density. 'B' represents the equation of revolving field moving in the positive α direction, and 'A' represent equation of revolving field moving in a positive direction. The field moving in the positive α direction is called the forward rotating field and in negative α direction is called the backward rotating field.

It is therefore concluded that a stationary pulsating magnetic field can be resolved due to two rotating magnetic fields both of equal magnitude and moving at synchronous speed in the opposite direction at the same frequency as the stationary magnetic field.

The theory based on such a resolution of an alternating field into two counter-rotating fields is called the **Double revolving** field theory of single phase induction machine.

CHAPTER-9: CONSTRUCTION OF THREE PHASE SYNCHRONOUS MACHINES

An alternator is composed of two main parts: the stator and the rotor. The stator remains stationary while the rotor rotates. Within the stator lies the armature winding, responsible for generating voltage, from which the output is derived. On the other hand, the rotor generates the primary flux essential for the alternator's operation.

Stator Construction

The stator comprises several key components, including the frame, stator core, stator windings, and cooling system. Typically, the frame is constructed from cast iron for smaller machines and welded steel for larger ones. To minimize losses due to hysteresis and eddy currents, the stator core is crafted using laminations of high-grade silicon steel. Within the stator, a three-phase winding is placed in slots along the inner periphery, with the winding being star-connected and distributed across multiple slots. When current flows through this distributed winding, it generates a nearly sinusoidal spatial distribution of electromotive force (e.m.f.).

Rotor Construction

The rotor construction is of two types

1. Salient-pole type.
2. Cylindrical rotor type.

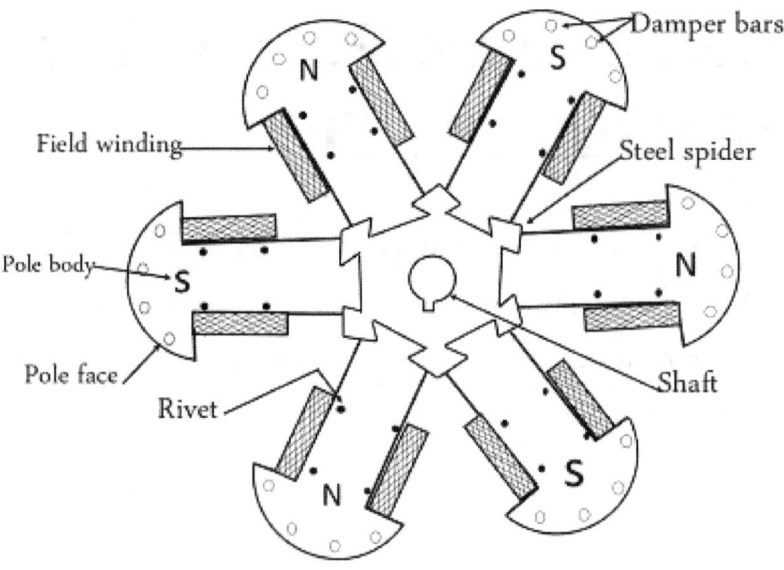

Figure: Six-pole salient-pole rotor

Salient-pole rotor

The term salient refers to 'projecting'. A salient-pole consists of poles that are projected out from the surface of the rotor core. These are used for the rotors for four or more poles.

The rotor is subjected to changing magnetic fields that is why it is made of steel laminations to reduce the eddy current losses. Identical dimensions poles are assembled by stacking laminations to the required length and then riveted together. After the field coil is placed around each pole body, these poles are fitted by a dove-tail joint to a steel spider keyed to the shaft. Salient-pole rotors have faces to damp out the rotor oscillations during a sudden change in load conditions. A non-uniform air gap accompanies a salient-pole synchronous machine.

The air gap is minimum under the pole centers, and it is maximum in between the poles. The pole faces are so shaped that the radical air gap length increases from the pole center to the pole tips so that the flux distribution in the air gap is sinusoidal. This will help the machine to generate sinusoidal e.m.f.

To give the alternate north and south polarities, the individual field-pole windings are connected in series. The end of the field windings is connected to a d.c winding by the brushes on the slip-rings.

The salient-pole generators have a large number of poles and lower operating speed. Salient-pole alternators that are driven by water turbines are called **hydro-alternators or hydro generators**.

Cylindrical Rotor

Cylindrical-rotor machines are also known as **non-salient pole rotor machine**. The rotor construction involves forming a sleek cylinder without any physical poles, unlike the salient-pole design. These rotors are crafted from solid forgings comprising high-grade nickel-chrome-molybdenum steel.

In about two-thirds of the rotor periphery, slots are cut at regular intervals and parallel to the shaft. The d.c field windings are connected in these slots. The winding is of distributed type. The unslotted portion of the rotor forms two pole faces. These machines have a small diameter and long axial length.

Such construction limits the centrifugal forces. Thus, the cylindrical rotors are useful in high-speed machines. Steam or gas turbines drive Cylindrical-rotor machines. Cylindrical-rotor synchronous generators are called turbo-alternators or turbo-generators.

CHAPTER-10: ELECTROMECHANICAL ENERGY CONVERSION PRINCIPLES

An electromechanical energy conversion device converts electrical energy into mechanical energy or, mechanical energy into electrical energy. Electromechanical energy conversion takes place via the medium of a magnetic field or an electric field, but most practical converters use magnetic field as the coupling medium between electrical and mechanical systems, this is because the electric storing capacity of the magnetic field is much higher than that of the electric field. Electromechanical energy converters are either gross-motion devices such as microphones, loudspeakers, electromagnetic relays, and certain electrical measuring instruments, etc.

DC, induction and synchronous machines are used extensively for electromechanical energy conversion. When the conversion takes place from electrical to mechanical form, the device is called the motor, and when the mechanical energy is converted to electrical energy, the device is called a generator. In these machines, conversion of energy from electrical to mechanical form or from mechanical to electrical from results from the following two electromagnetic phenomena:

1. When a conductor is allowed to move in a magnetic field, a voltage is induced in the conductor.
2. When a current-carrying conductor is placed in a magnetic field, then a mechanical force is experienced by the conductor.

During the monitoring process, current passes through conductors positioned within a magnetic field, generating a force on each conductor. These conductors are situated on the rotor, which can freely move. The resulting electromagnetic torque on the rotor is then transmitted to the rotor shaft, potentially driving a mechanical load. As the conductors rotate within the magnetic field, a voltage is induced in each conductor. Conversely, in the generating process, the rotor is driven by a prime mover, inducing a voltage in the rotor conductors. When an electrical load is connected to the winding formed by these conductors, a current flows, supplying electric power to the load. Additionally, the current passing through the conductors interacts with the magnetic field, generating a reaction torque that opposes the torque produced by the prime mover.

Conservation of energy

According to the principle of conservation of energy, energy can neither be created nor be destroyed it can only be transformed from one state to another.

In an energy conversion device, the total input energy is equal to the sum of the following three components:

Thus, with an electromechanical conversion device, the energy balance equation can be written as

$$\begin{bmatrix} \text{Electrical} \\ \text{energy} \\ \text{input} \end{bmatrix} = \begin{bmatrix} \text{Energy to} \\ \text{electrical} \\ \text{losses} \end{bmatrix} = \begin{bmatrix} \text{Energy to field} \\ \text{storage in the} \\ \text{electrical system} \end{bmatrix} = \begin{bmatrix} \text{Mechanical} \\ \text{energy} \\ \text{output} \end{bmatrix}$$

The above equation is for motor action. For generator action, the energy balance equation is written as

$$\begin{bmatrix} \text{Total mechanical} \\ \text{energy input} \end{bmatrix} = \begin{bmatrix} \text{Electrical energy} \\ \text{output} \end{bmatrix} + \begin{bmatrix} \text{Total energy} \\ \text{stored} \end{bmatrix} + \begin{bmatrix} \text{Total energy} \\ \text{dissipated} \end{bmatrix}$$

CHAPTER-11: ARMATURE REACTION IN DC GENERATOR

DC Generator

A DC generator is a device which converts mechanical energy into DC electrical power via electromagnetic induction.

Whenever there's a change in the magnetic flux connecting a conductor, it generates an electromagnetic field (EMF), which is the underlying principle behind the operation of a DC generator. In a DC generator, both field and armature windings play crucial roles. The alternating EMF generated in the armature winding is converted into direct voltage by a commutator, which is fixed on the generator shaft. While the field winding is typically located on the stator, the armature winding is positioned on the rotor.

Working of a DC Generator

As displayed in the below picture, a single turn loop 'ABCD' in a single loop DC generator rotates clockwise in a consistent magnetic field at a constant speed. The magnetic flux connecting the coil sides "AB" and "CD" continually varies as the loop revolves. The electromotive Force in one coil side is added to the electromotive Force in the other as a result of the change in flux linkage.

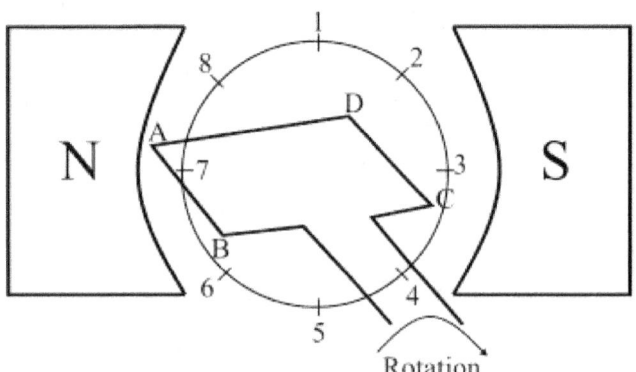
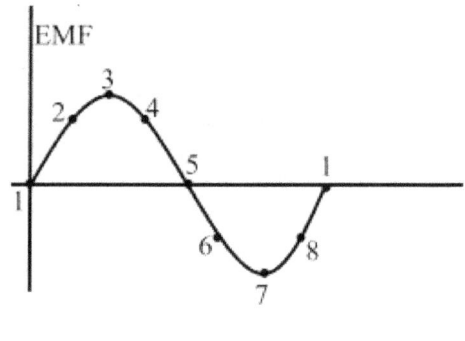

The following describes how EMF is produced by a DC generator:

- Because the motion of the coil sides is parallel to the magnetic flux while the loop is in position 1, there is no EMF produced.
- Due to the coil sides spinning at an angle to the magnetic flux while the loop is in position 2, there is a little EMF.
- The coil sides move at a 90^0 to the magnetic flux while the loop is in position 3, generating the most EMF.
- The coil sides create less EMF when the loop is in position 4 because they are cutting the magnetic flux at an angle.
- When the loop is in position 5, it is moving parallel to the magnetic flux and there is no flux coupling with the coil side. The coil therefore doesn't generate any EMF.
- Position 6 is a pole with an opposite polarity, behind which the coil sides pass, reversing the polarity of the generated EMF. The biggest EMF will be produced in this direction at point 7, whereas there will be no EMF at position 1. This cycle repeats with every coil turn.

It is obvious that the loop's produced EMF is an alternating one. This is such that any coil side, like AB, has EMF in one way when affected by N-pole while in the other way when affected by S-pole. As a result, an alternating current can flow through with a load when it is linked across the generator's terminals. This *alternating emf produced in the loop may now be changed into direct voltage by utilizing a commutator*.

Armature Reaction

The term "armature flux" refers to the magnetic field created by the current flowing through the armature conductors. This armature flux distorts and reduces the magnetic flux produced by the primary poles. ***This interaction between the armature flux and the main flux is referred to as armature response.***

1. Case1:-

Let's consider a two-pole generator which is idle. Therefore, there is no current flowing through the armature conductors. Only the major flux (φ_m), which is generated by the main poles, is present in the machine in this situation. The distribution of this primary flow with respect to the polar axis is symmetrical (i.e. center line of field poles).

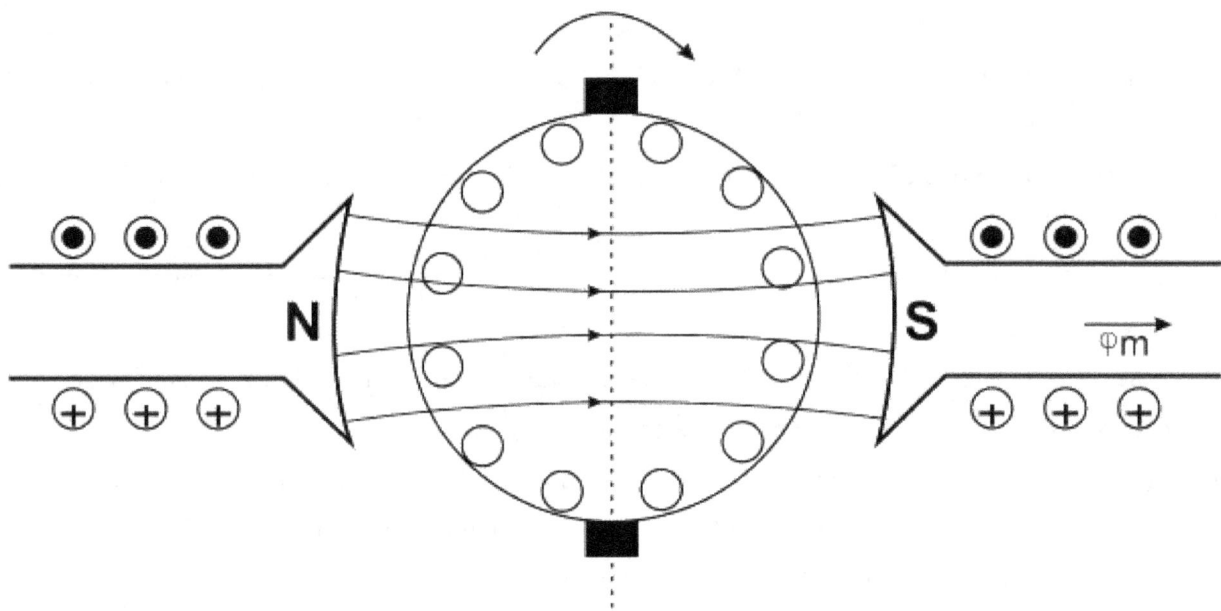

The **magnetic neutral axis** (MNA), which is a plane perpendicular to the flux axis, coincides with the geometrical neutral axis. The MNA is also known as the axis of commutation since the brushes are always positioned along it.

2. Case2:

Let's consider an armature that is operating with no current flowing through the field coils. To determine the direction of flux produced by the current flowing through the armature conductors, use the cork-screw rule. Referring to the illustration, current flows into the paper plane through the conductors beneath the N-pole. As a result, the conductors beneath the N-pole are producing a downward flux.

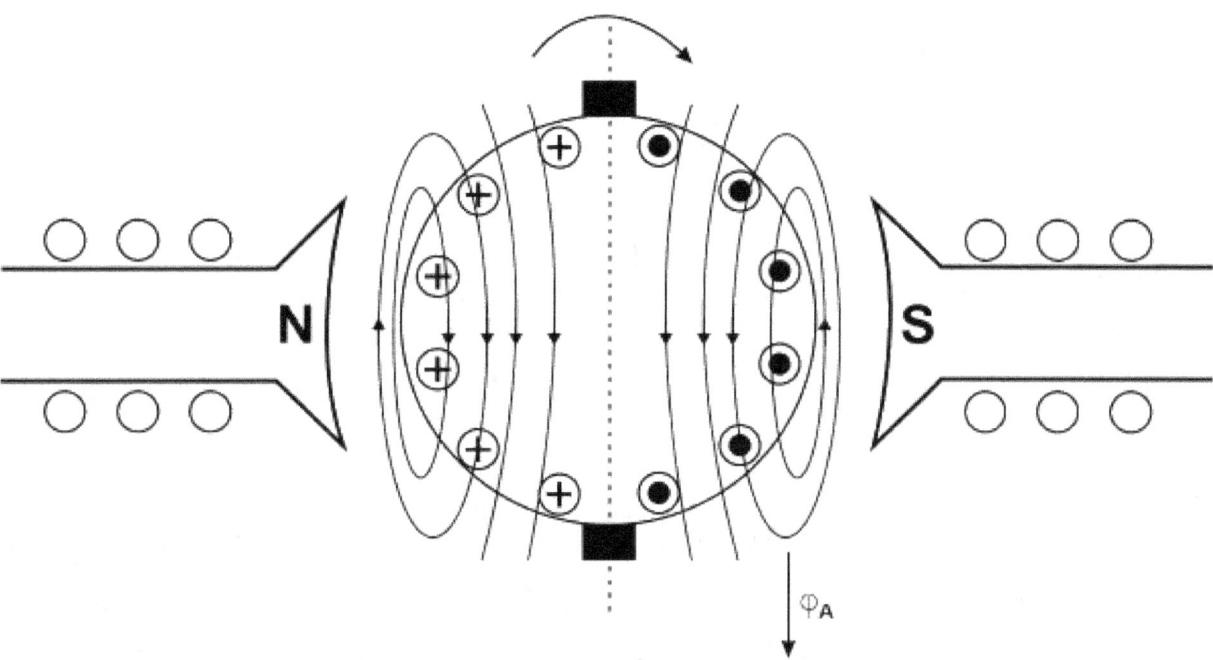

Comparably, the conductors under the S-pole transport current away from the paper's surface. Additionally, a downward-directed flux is produced by these conductors. As a result, a downward flux is produced via the armature by the whole armature conductor. Armature flux (φ_A) is the name given to this flux.

3. Case3:-

In this case, the simultaneous action of the armature and field currents is demonstrated. As a result, the machine has two fluxes inside it, one created by the generator's main field poles and the other by the current flowing through the armature conductors. The resultant flux (φ_R) is created by combining these two fluxes.

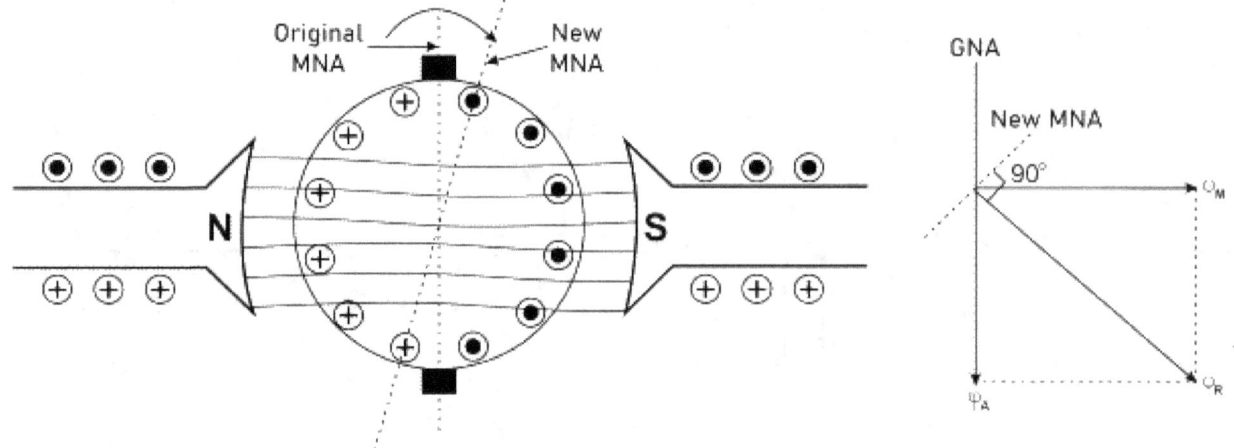

The principal flux hitting the armature is shown to be displaced and twisted from the explanation above. The distortion causes an increase in flux density at both the lower pole tip of the S-pole and the higher pole tip of the N-pole. At the lower pole tip of the N-pole and the higher pole tip of the S-pole, the flux density is also declining. As a result, the path of the resulting flux has changed to coincide with the generator's rotational direction.

The MNA is also displaced since it is always perpendicular to the axis of the resulting flux. The rise in flux in one pole tip is less than the reduction in flux in the other pole tip because of the non-linear behavior and saturation of the core. The major flow is thus reduced. When a result, as the load increases, the produced emf ($E_g \propto N\varphi_m$) decreases.

Armature reaction effects

The armature reaction in a DC generator results in the following detrimental effects.

1. The generated EMF is decreased as a result of a modest reduction in the overall field flux created by each pole.
2. The MNA is likewise displaced in the generator's rotational direction as a result of the moving of the resulting flux axis.
3. The armature response causes a flux to form in the neutral or commutating zone. The conductor voltage that is brought on by this flux in the neutral zone is what interferes with commutation.

The Armature Reaction Effect: Solutions

The armature response issue may be minimized using different techniques.

1. Position the brush differently-

Using this technique, turn the brush mechanism until it is positioned correctly in the neutral zone. Only current with a fixed load can be used for this.

2. The poles' ends should be changed-

In this procedure, the field pole tip must be altered so that the high reluctance channel prevents high flux from occurring on the ends.

3. Interpoles-

The impact of armature response can be reduced by sandwiching a series of interpoles or commutating poles between the DC generator's main poles. The interpole's polarity must match that of the main pole directly to its left in the rotational direction. In order for the associated fluxes to fluctuate together with the load current, the interpole windings and armature are connected in series.

4. Compensating Winding

The armature reaction varies dramatically as a result of the heavy load activities. The armature flux in these generators is not sufficiently neutralized by the interpoles. Therefore, compensatory windings are employed to solve this issue.

An auxiliary winding that is inserted into the slots of the main poles is the compensating winding. The compensatory winding and armature are linked in series such that the current flowing through any given pole face of the compensating conductors will be flowing in the opposite direction to that of the neighboring armature conductors. As a result, the compensatory windings provide a flux that is equal to and in the opposite direction of the armature flux, totally canceling out the armature response.

5. By Increasing Brush-Contact Resistance :

By using a strong brush contact resistance with the commutator part that is being commutated, sparking can be avoided. Let's examine the process of doing it.

Imagine a brush interacting with segment 1 of a winding coil "A" having two commutator segments. When seen in figure (a) below, as segment 2 gets closer to the brush, a portion of the current I passes through coil "A" in a forward direction.

. In figure (b), no current flows in coil "A" while the brush is between the two segments. The stream now flows in the other direction as segment 1 leaves the bush in the illustration (c).

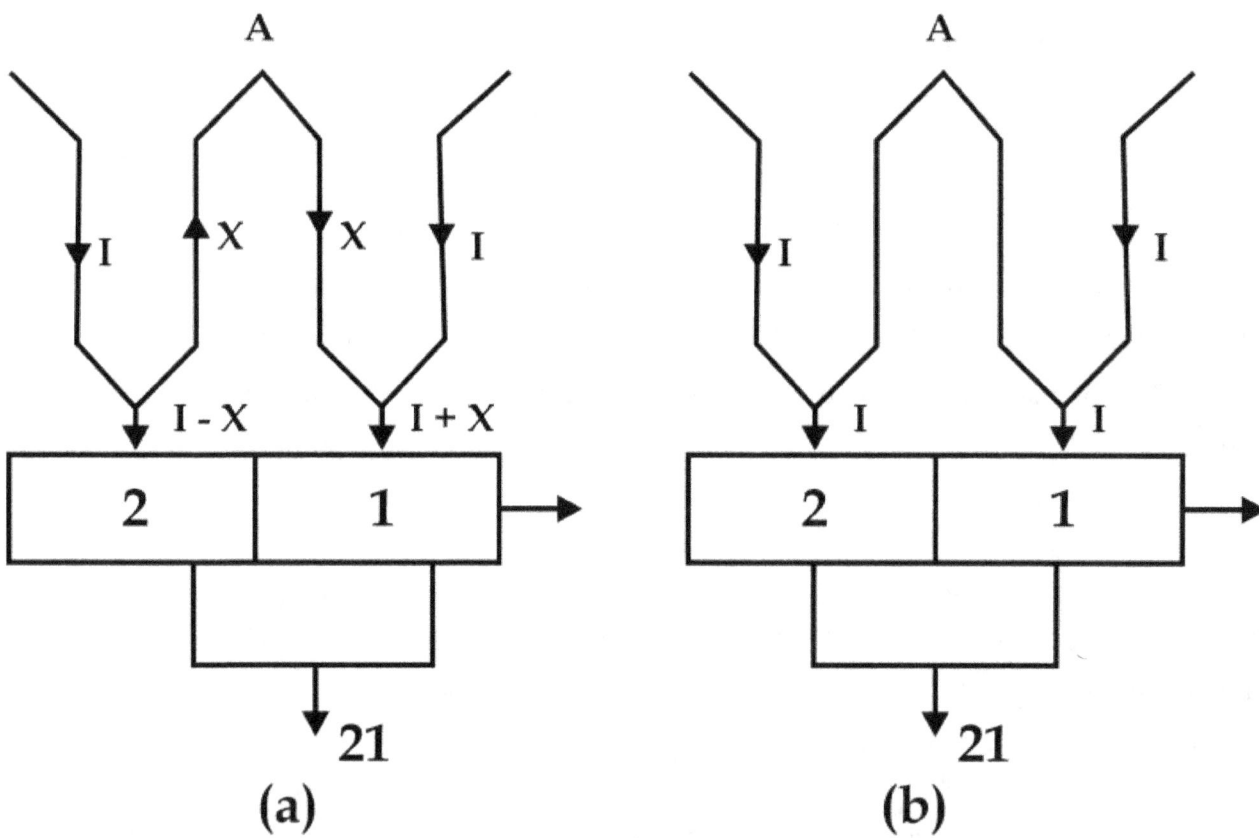

As segment 2 completes the process and contacts the brush, coil A experiences an increase in current flow. In order for the current in coil A to entirely reverse by the time the brush has reached segment 2, as seen in figure (d). Sparking is created on the surface of segment 1 with the brush during this full reversal of current.

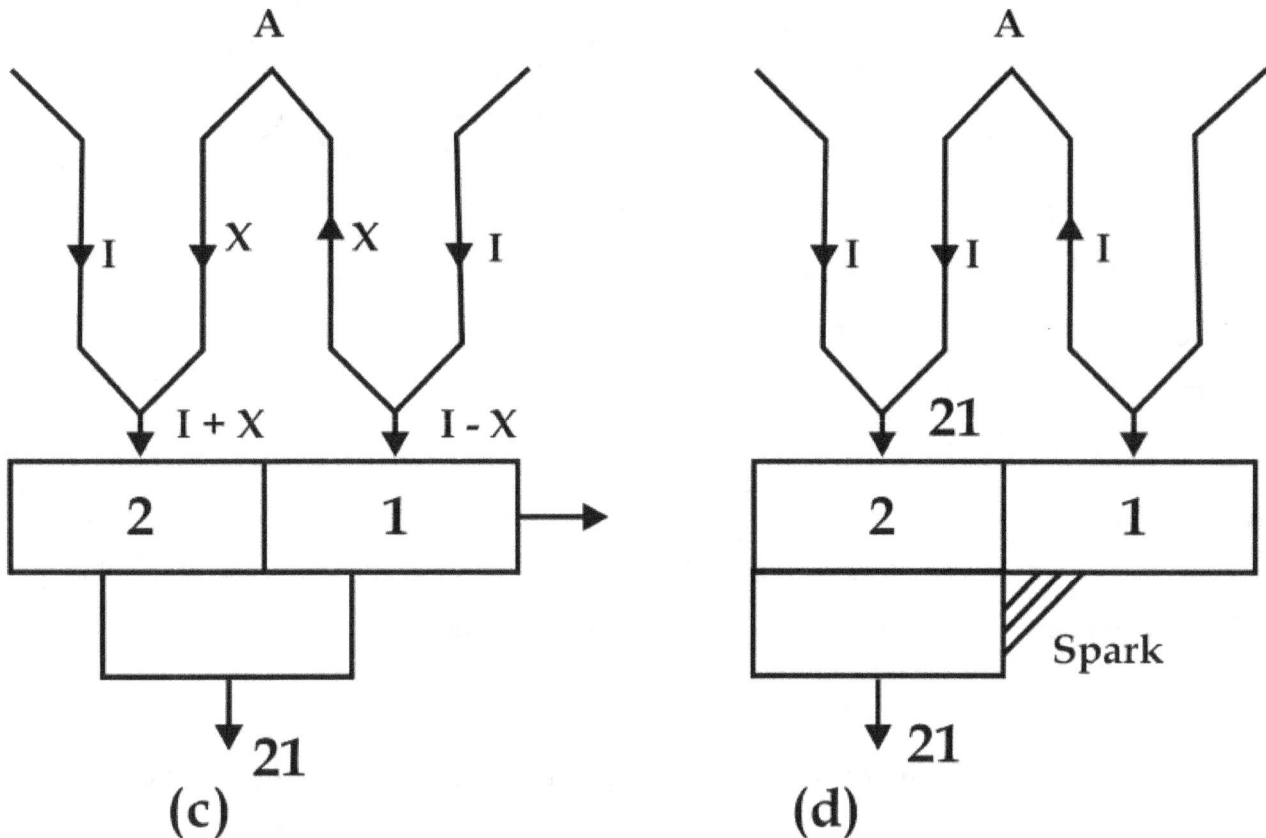

Thus, by increasing the brush-contact resistance (using carbon brushes with segments), sparking during current reversal may be prevented. The contact resistance is arranged such that the brush has a high resistance and the commutator segment has a reduced area of contact (segment 1 in the example in figure (c) above).

CHAPTER-12: ARMATURE REACTION IN SYNCHRONOUS GENERATOR

Synchronous Generator

A synchronous generator is a device that uses electromagnetic induction to transform mechanical power into alternating current (AC) electricity.

Alternators and *AC generators* are other names for synchronous generators. Since it generates AC power, it is referred to as an "alternator." Because it has to be run at synchronous speed in order to generate AC electricity at the required frequency, this generator is known as a synchronous generator.

There are two types of synchronous generators: **single-phase and poly-phase (generally 3 phase).**

Working Principle and Operation of Alternator

An electromagnetic field (EMF) is formed in a conductor when the flux connecting to it changes. This is the way an alternator or synchronous generator operates. When the alternator's armature winding is exposed to a revolving magnetic field, the winding will produce voltage.

The alternate N and S poles develop on the rotor when the alternator's rotor field winding is powered by a DC exciter. The magnetic field of the rotor poles cuts the armature conductors mounted on the stator as the rotor is turned counterclockwise by a primary mover.

As a result, electromagnetic induction induces the EMF in the armature conductors. Because the rotor's N and S poles pass the armature conductors alternately, the induced EMF is alternating.

The **Fleming's right-hand rule** may be used to identify the path of the produced EMF, and the frequency is found by

1. f= NSP/120

Where,

The synchronous speed in RP is **Ns**.

P is the rotor poles' total number.

The rotor's rotational speed as well as the DC field excitation current both affect how much voltage is produced. When the winding is balanced, the voltage generated in each phase is the same yet differs electrically by 120° in phase.

Armature Reaction in Synchronous Generator

On the basis of Faraday's law, every spinning electrical machine operates. Every electrical device needs a magnetic field, a coil (known as an armature), and a motion in relation to each other. When using an alternator, we provide electricity to the pole to create a magnetic field, and the armature provides the output power. The conductors of the armatures cut the magnetic field's flux due to the relative motion between the field and the armature, causing a changing flux linkage with these conductors. The armature would experience an electromagnetic field (EMF) in accordance with Faraday's law of electromagnetic induction. As a result, the armature coil starts to conduct current as soon as the load is attached to the armature terminals.

As soon as current begins to flow through the armature conductor, it has one reversible influence on the synchronous generator main field flux (or alternator). The **armature response** in an alternator or synchronous generator is the technical term for this adverse impact. In other words, *armature response refers to the interaction between armature (stator) flux and the flux generated by the rotor field poles.*

We obviously know that a current-carrying conductor generates one's own magnetic field, and that this magnetic field has an impact on the alternator's primary magnetic field.

Armature Reaction in Alternators

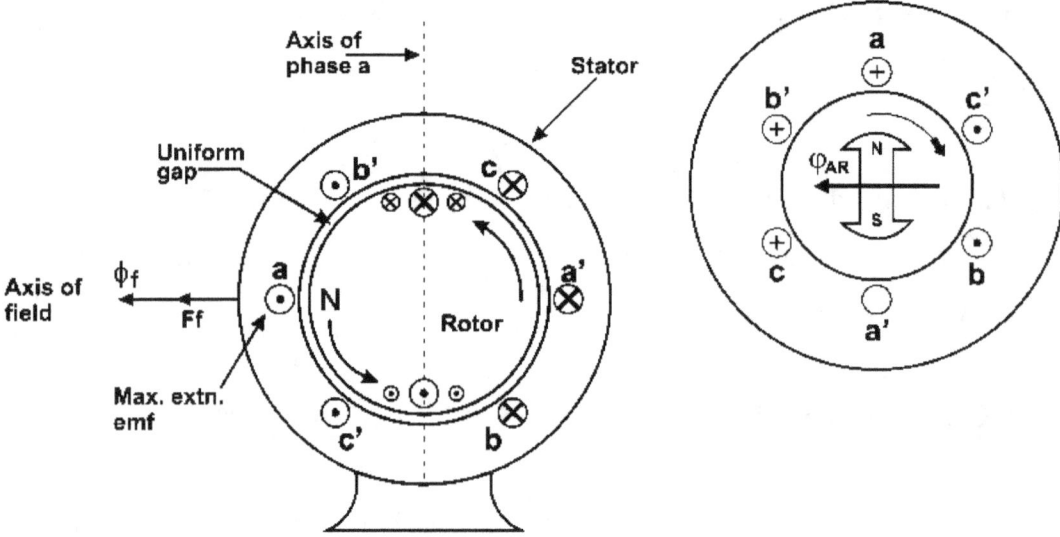

Either the main field is distorted, the main field flux is decreased, or both of these unfavorable consequences occur. They cause the machine's performance to decline. There is a phenomenon called the cross-magnetizing effect when the field is deformed. And the demagnetizing effect is what happens as the field flow decreases.

The magnetic field acts as a channel for the electromechanical energy conversion. An emf, whose magnitude depends on the relative speed and magnetic flux, is produced in the armature windings as a result of the relative motion between the armature conductors and the main field. The armature reaction causes flux to be reduced or distorted, which in turn affects the net induced emf and lowers the machine's performance.

Now let us see Armature Reaction in Alternator

The impact of armature response in an alternator, like that in all other synchronous machines, depends on the power factor, or the phase connection between both the armature current and terminal voltage.

When a generator powers a lagging load, it is delivering magnetic energy to the load because reactive power (also known as lagging) is the magnetic field energy. The generator's net reactive power decreases since this power is derived from synchronous machine stimulation.

The armature response is demagnetizing as a result. Similar to this, the generator supplies a leading load (as the leading load absorbs the leading VAR) and in exchange receives trailing VAR (magnetic energy) from the generator, which has a magnetizing effect. The only armature reaction when the load is solely resistive is cross magnetization.

The phase angle between the stator armature current and the induced voltage across the alternator's armature winding determines the armature response of an alternator or synchronous generator.

Armature current and voltage have a phase mismatch between them that can range from -90 degrees to +90 degrees.

Therefore we can say that angle lies between -90° to + 90°.

We will look at three typical scenarios in order to comprehend the true impact of this angle on the alternator's armature reaction,

- **When θ = 0**
- **When θ = +90°**
- **When θ = - 90°**

1. Armature Reaction of Synchronous Generator at Unity Power Factor

The angle between the armature current I and the induced emf E is zero at unity power factor. This indicates that the armature current and the induced emf are in phase. Theoretically, however, we are aware that the armature's induced emf results from a fluctuating main field flux that is connected to the armature conductor.

The main field flux remains constant in relation to the field magnets because the field is stimulated by DC, but it would alternate in relation to the armature since the field and armature are moving relative to one another in the alternator. If the alternator's primary field flux in relation to the armature may be expressed as

$\Phi_f = \Phi_{fm} \sin wt \ldots\ldots\ldots(1)$

The induced emf across the armature is thus $d\Phi_f/dt$ proportional.

Now $d\Phi_f/dt = -w\Phi_{fm}\cos wt \ldots\ldots(2)$

Hence, from these above equations (1) and (2), it is clear that the angle between, φf and the induced emf E will be 90°.

Armature flux an is now inversely correlated with armature current I. As a result, armature flux Φa and armature current I are in phase.

I and E are in phase at unity electrical power factor once more. As a result, at unity power factor, a and E are in phase. Therefore, at this point, the field flux and induced emf E are in quadrature, and the armature flux is now in phase with E. As a result, the armature flux a and the main field flux f are in quadrature.

Due to the two fluxes' perpendicular relationship, the alternator's armature response at unity power factor is solely of the distorting or cross-magnetizing kind.

The arrangement of main field flux under a pole face doesn't really stay uniform as the armature flux drives the main field flux perpendicularly. Beneath the leading pole tips, the flux density falls while it increases somewhat under the following pole tips.

2. Armature Reaction of Alternator at Lagging Zero Power Factor

The armature current lags the induced emf in the armature by 90 degrees at trailing zero electrical power factor.

Since the main field flux induces an emf in the armature coil, the emf advances the main field flux by 90 degrees. The field flow is obtained from equation (1).

$$\Phi_f = \Phi_{fm} \sin wt$$

Therefore, induced emf $E \propto -d\Phi_f/dt$

$$E \propto -w\Phi_{fm}\cos wt$$

Thus, E is at its maximum and Φ_f is zero at time wt = 0.

E is zero and Φ_f is at its highest value at wt = 90°.

E is at its maximum and Φ_f is zero when wt = 180°.

E is zero and Φ_f has a maximum negative value at wt = 270°.

Here, Φ_f achieved its maximum value before E. So, f is 90 degrees ahead of E.

Armature current I now lags E by 90 degrees and is proportional to armature flux a. Thus, Φ_a lags E by 90 degrees.

As a result, armature flux and the field flux have an exact opposite effect. As a result, the alternator's armature reaction at lagging zero power factor is wholly demagnetizing. This implies that armature flux impedes main field flow directly.

3. Armature Reaction of Alternator at Leading Power Factor

At conditions of leading power factor, armature current "I" trails induced emf "E" by a 90° angle. Once more, we've demonstrated that field flux f leads to induced emf E via a 90° angle.

Once more, armature flux an is inversely correlated with armature current I. As a result, Φ_a and I are in phase. As a result, armature flux Φ_a leads E by 90 degrees, just as I do. Field flux and armature flux are considered to be in the same direction in this example since they both induce emf E at a 90° angle. As a result, the combined flux of the armature and field is easily calculated.

Consequently, it may be concluded that the alternator armature response caused by a wholly leading electrical power factor is of the magnetizing kind.

Nature of Armature Reaction

1. The armature response flux spins at synchronous speed and has a constant magnitude.
2. When the generator produces a load at unity power factor, the armature reaction is cross-magnetizing.
3. When the generator is operating at the leading power factor and serving a load, the armature reaction is partially demagnetizing and partially cross-magnetizing.
4. The generator's armature reaction is partially magnetizing and partially cross-magnetizing when it serves a load at a leading power factor.
5. The main field flux does not affect the armature flux.

CHAPTER-13: ARMATURE REACTION IN DC MACHINE

What is a DC machine?

Electrical energy can be converted into mechanical energy or the other way around using an electromechanical device known as a **DC machine.**

The DC machine, sometimes referred to as the DC motor or DC generator, converts electrical energy into mechanical energy and mechanical energy into electrical energy, respectively. The same device can function as a generator or a motor. Both the DC generator and the DC motor have the same architecture.

Working of DC machine

The basis for the operation of a DC machine is the torque produced when a current-carrying conductor coil is placed in a magnetic field. This torque causes the conductor coil to rotate inside the magnetic field. To determine the direction of this created torque, use the **Fleming's left-hand rule**. Following is the computed generated force.

1. F=BIL

Where,

F = Magnitude of the generated force

- **B = Flux density**
- **I = Current**
- **L = the length of the conductor**

Armature Reaction in DC Machine

The carbon brushes are always positioned at the magnetic neutral axis in a DC machine. The geometric neutral axis and the electromagnetic neutral axis meet in a no-load state. Therefore, when the machine is loaded, its armature flux has a triangle wave structure and is directed along the interpolar axis (the axis between the magnetic poles). The main field becomes cross-magnetized as a result, and the brush axis becomes the direction of the armature current flow. Due to the cross-magnetization effect, flux is concentrated at the leading pole tip while a motor is operating and on the trailing pole tip when a generator is operating.

The term "armature reaction" refers to how the armature flux affects the main flux. A DC motor's resultant flux is stronger at the tips of the leading pole and weaker at the tips of the following pole.

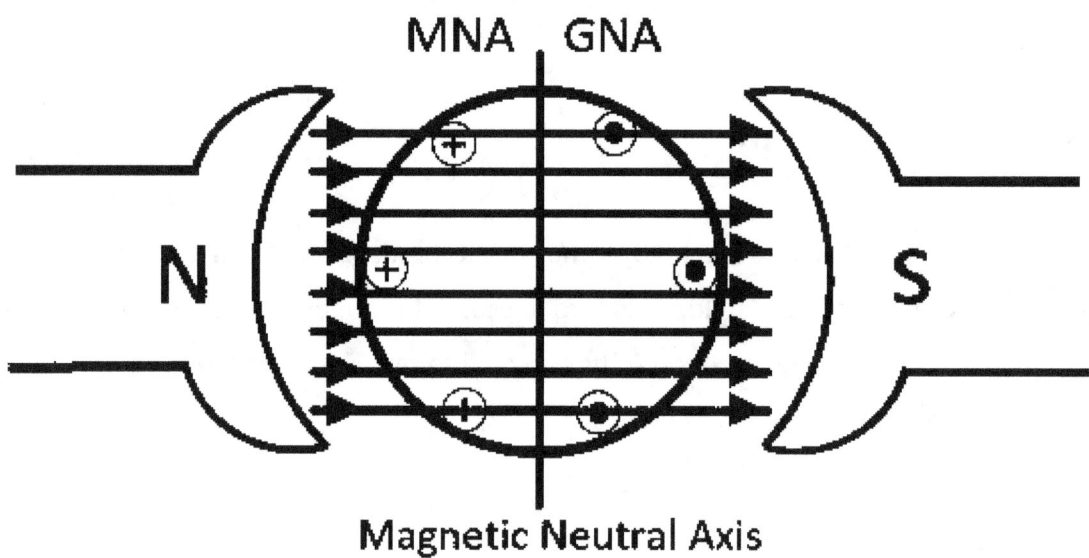

What do Leading and Trailing Pole Tips mean?

The leading tip of a pole is where the armature conductors enter the magnetic influence, while the trailing tip is situated in the opposite direction. For example, in the scenario where the motor rotates counterclockwise, the lower tip of the North Pole serves as the leading tip, while the higher tip of the South Pole acts as the leading tip. However, these designations are reversed if the motion is reversed, as in the case of a generator. Due to cross magnetization, the magnetic neutral axis shifts under load along the direction of rotation in a DC generator and in the opposite direction in a DC motor. If the brushes are not repositioned accordingly, the generated e.m.f. in the generator or motor decreases, resulting in significant sparking during commutation. This occurs because only the brushes' coils are subject to commutation, and the alternate pole influences the commutating coil, shifting its location from the north to south pole or vice versa. As a result, the direction of the current quickly switches from +i to -i or vice versa. As a result, the coil encounters a very high reactance voltage (L* di/dt), which destroys the brushes and commutator section by escaping as heat energy and sparking. The following techniques are employed to lessen the aforementioned negative impacts and enhance the functioning of the machine:

Brush Shift

The brushes can move in a way which reduces the air gap between the flux, such as all along direction of rotation for generator action and opposite the direction of rotation for motor action. This will raise the speed of the motor and lower the induced voltage in the generator. The demagnetizing magneto motive force (mmf) resulting from this is provided by:

$F = NI$

Where, N - numbers of turns of the inductive coil

I - current

Brush shift is severely constrained; therefore, the brushes must be moved to a new location if the load, the rotation's direction, or the mode of operation changes. Due to this, only extremely tiny machines may use brush shift. The brushes are likewise fixed in this location at a position that corresponds to the regular load and mode of operation. These drawbacks make this strategy less popular in general.

Inter Pole

Almost all medium and big sized DC machines now utilize inter poles due to the brush shift's limitations. The inter polar axis has long, thin poles called inter poles. In the case of generator action, the succeeding pole is the one that will be rotated next, while in the case of motor action, the following pole is the one that will be rotated after the generator action. The inter pole's purpose is to balance the inter polar axis's armature response mmf. Since inter poles and the armature are connected in series, the direction of the inter pole changes when the armature's current flows in a different direction.

This is as a result of the armature reaction's mmf direction being in the interpolar axis. The reactance voltage (**L di/dt**) is totally neutralized by the commutation voltage, which is also provided for the coil that is undergoing commutation. As a result, sparking is prevented.

Inter polar windings carry the armature current and operate as intended regardless of load, rotational direction, or mode of operation because they are always kept in series with the armature. To guarantee that they exclusively affect the coil that is undergoing commutation and that their impact does not extend to the other coils, inter poles are made smaller. To prevent saturation and enhance responsiveness, the base of the interpoles is made broader.

Compensating Winding

The commutation difficulty with DC devices is not the sole issue. When operating with significant loads, the cross-magnetizing armature response may result in extremely high flux densities at the leading pole tip and trailing pole tip of the generator and motor, respectively.

This coil may produce an induced voltage high enough to cause a flashover between the related neighboring commutator segments because it is situated close to the commutation zone (at the brushes), where the air temperature may already be high as a result of the commutation process.

This flashover might affect nearby commutator segments and eventually ignite a full-scale fire that spreads from brush to brush over the commutator surface. The voltage L* di/dt that appears across the adjacent commutator segments of the machine may also increase to a value high enough to result in flash over between the adjacent commutator segments when the machine is subjected to rapidly changing loads. As the coil underneath it has the greatest inductance, this would begin at the pole's core. This might result in a fire similar to the one previously mentioned. This issue is more severe when the load is shifting from generating to motor action, as the induced e.m.f. and voltage L di/dt will then support one another.

The compensating winding is made up of parallel-to-the-shaft conductors implanted in the pole face that carry an armature current in the opposite direction as the armature conductors beneath that pole arc. The primary field has been fully compensated. Inductor in the armature circuit is also reduced, which enhances system responsiveness. No matter the load, rotational direction, or mode of operation, compensating winding performs adequately. Naturally, it aids in commutation since the inter polar winding is relieved of its responsibility to offset the armature mmf beneath the pole arc.

Major problems of compensating windings:

1. In huge machines that are plugged or heavily overloaded
2. Small motors are vulnerable to abrupt reversal and rapid acceleration.

Key Points

1. Armature conductors positioned beneath the pole arc primarily induce the cross-magnetizing armature reaction effect. This effect may lead to an increased flux density at the leading and trailing pole tips of a generator under heavy loads (or in a motor). However, due to saturation in the pole shoe, the increase in flux density might be less than the reduction in flux density in the remaining part of the pole shoe. Consequently, there could be a net decrease in flux per pole, often referred to as the demagnetizing effect of cross-magnetizing armature reaction, attributed to compensating windings.
2. The armature winding is interconnected in series with compensatory and interpolar windings situated on opposite sides of the armature.
3. The primary function of the interpolar winding is to optimize the commutation process, while the compensating winding aims to stabilize the net air gap flux, ensuring it remains constant despite variations.

CHAPTER-14: WHAT IS AN AC SERVO MOTOR?

The term "AC servo motor" denotes a specific type of servo motor utilizing AC electrical input to produce precise angular velocity as mechanical output. Despite some engineering distinctions, AC servo motors are essentially two-phase induction motors. These motors can provide output power ranging from a few watts to several hundred watts, with a usable frequency range typically spanning from 50 to 400 Hz. Additionally, employing a specific type of encoder to relay information about speed and position enables closed-loop control within the feedback system.

Construction of an AC Servo Motor

While an AC servo motor shares the foundation of a two-phase induction motor, it incorporates unique design elements absent in standard induction motors. These distinct features contribute to the perception that AC servo motors are constructed quite differently from their conventional counterparts.

The stator and rotor are its two key structural components.

- **Stator:** To begin, look at the following illustration of an ac servomotor's stator:

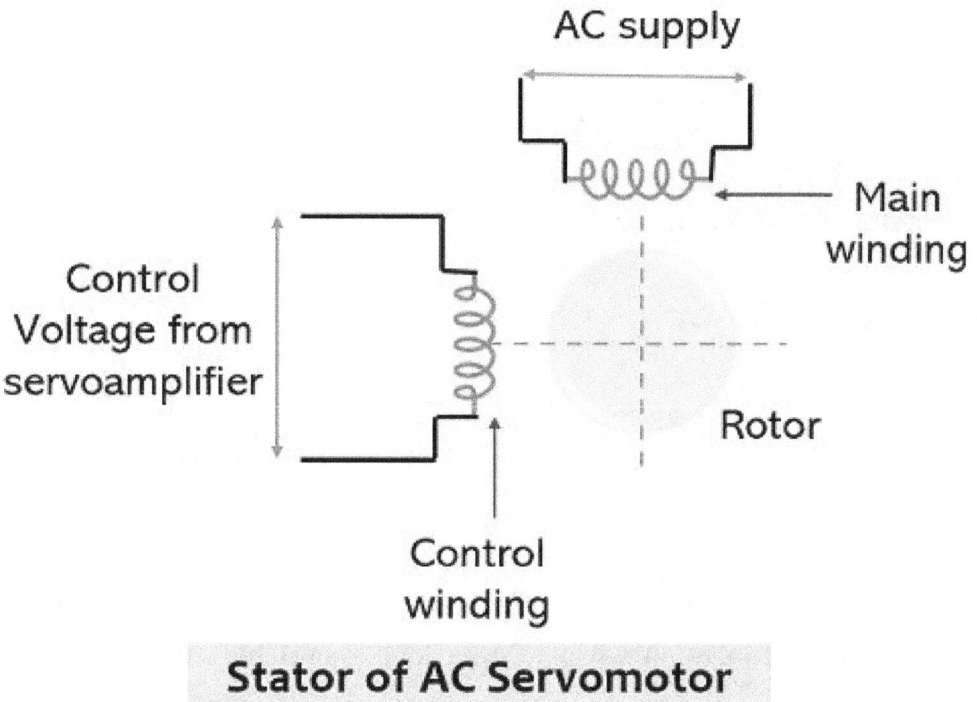

Stator of AC Servomotor

Two distinct windings that are evenly spaced apart and at 90° angles make up the stator of an ac servo motor. The term "main" or "fixed" winding refers to one of the two windings, while the term "control" refers to the other.

The primary winding of the stator receives an input that is a continuous ac signal. although the control winding receives variable control voltage, as the name indicates. The servo amplifier is used to produce this changing control voltage.

It should be noted that the voltage provided to the control winding must be 90 degrees out of phase with the input ac voltage in order to create a revolving magnetic field.

- **Rotor:** There are typically two sorts of rotors; one is a drag cup type and the other is a squirrel cage type.

Below is an illustration of a rotor with a squirrel cage:

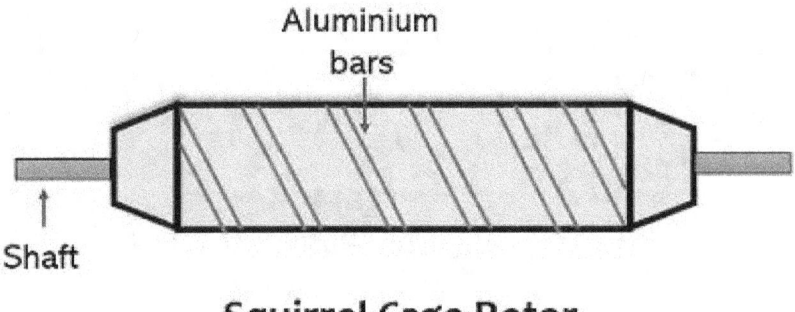

Squirrel Cage Rotor

This sort of rotor has a long, narrow diameter and is made of aluminum conductors, which makes it lighter in weight.

It should be noted that there are regions with both positive and negative slopes in the torque-speed characteristics of a conventional induction motor, which correspond to unstable and stable zones, respectively.

However, because ac servo motors are made to be highly stable, a positive slip zone cannot exist in their torque-slip characteristics. Additionally, the motor's produced torque must linearly decrease with speed.

The rotor circuit resistance should be high and have little inertia in order to do this. This is why it is important to have a reduced diameter to length equal while building the rotor.

The squirrel cage motor's smaller air gaps between its aluminum bars make it possible to use less magnetizing current.

Here is an illustration of a drag cup style rotor:

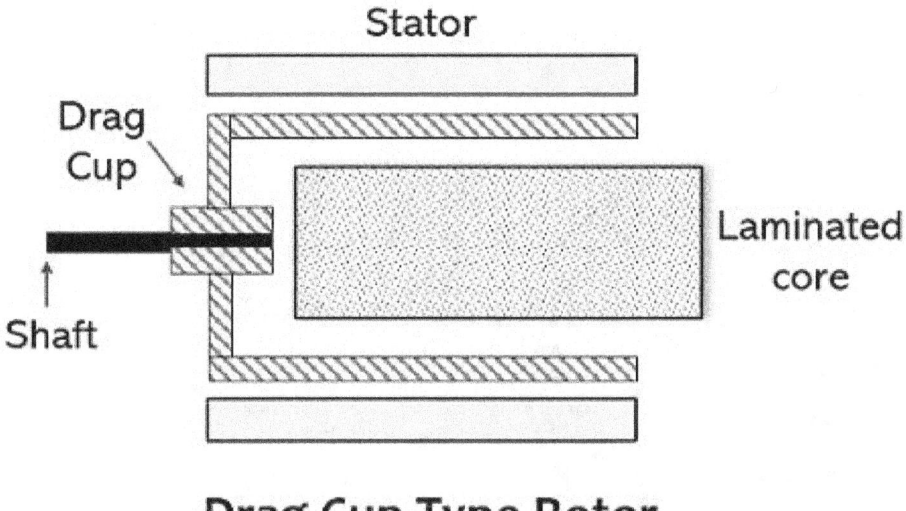

Drag Cup Type Rotor

The construction of this type of rotor is different from the squirrel cage type. It consists of a drag cup surrounded by an aluminum laminated core with small air holes on either side.

A drive shaft that is linked to these drag cups makes operating easier.

In applications where there is a minimal power need, the two air gaps on either side of the core reduce inertia.

Two Phase AC Servo Motor

Servo motors are available for two- and three-phase AC systems. The two dispersed windings in the stator of the two-phase AC servo motor are electrically 90 degrees apart from one another. One winding, referred to as the Reference or Fixed Phase, receives power from a source of constant voltage. The other one is called Control Phase, and it comes with a variable voltage supply.

Below is a graphic showing how the two-phase AC servo motor is connected:

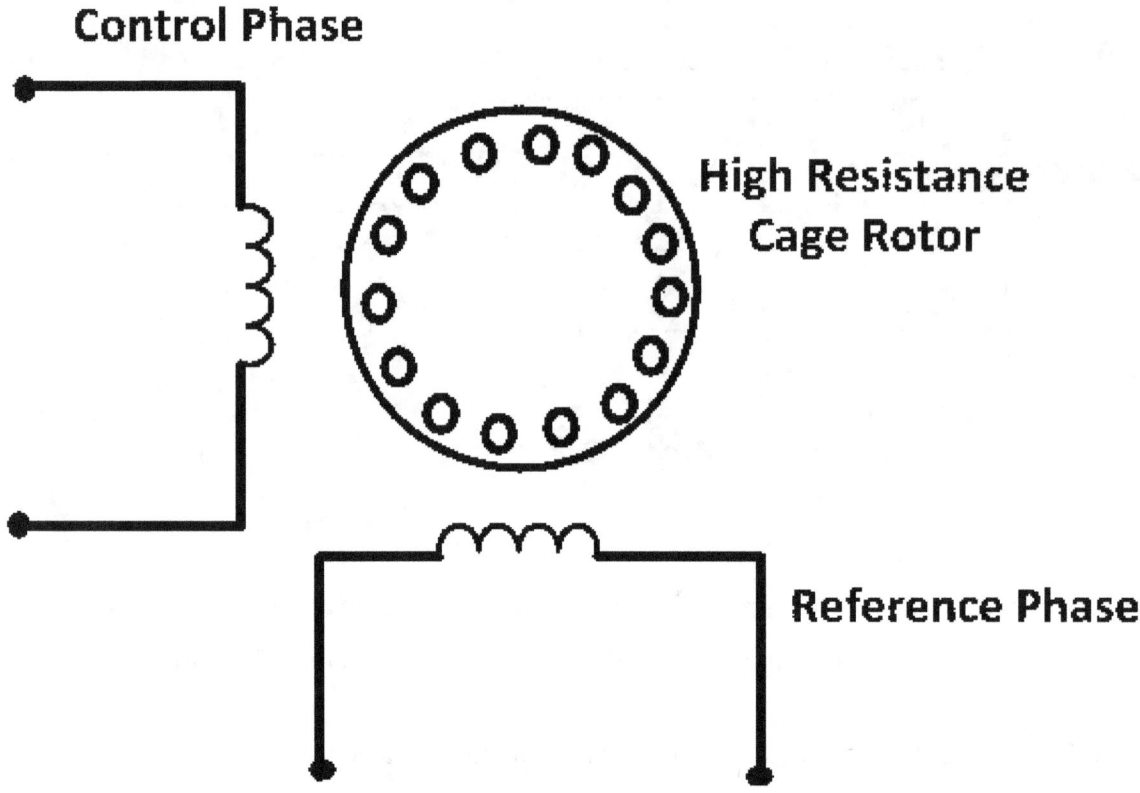

A servo amplifier is often used to supply the control phase. Phase difference between control voltage and reference phase voltage determines how fast and how much torque the rotor spins at. The phase difference can be switched from leading to lagging or vice versa to change the direction of the rotor's spinning.

The graphic below depicts the torque-speed characteristic of a two-phase AC servo motor.

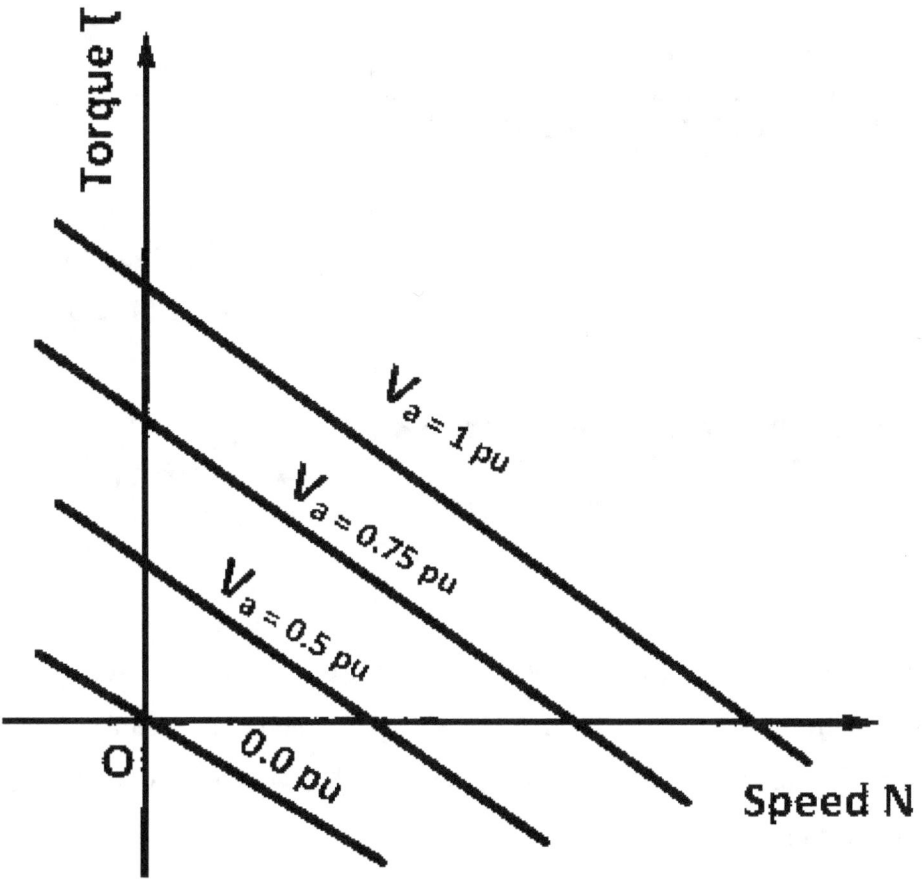

The motor takes positive damping from the negative slope, which indicates a high rotor resistance and improves stability. For practically all control voltages, the curve is linear. Based on information as the Drag Cup Servo motor, as illustrated in the image below, the weight and inertia of the motor are reduced to increase the motor's reaction to a light control input.

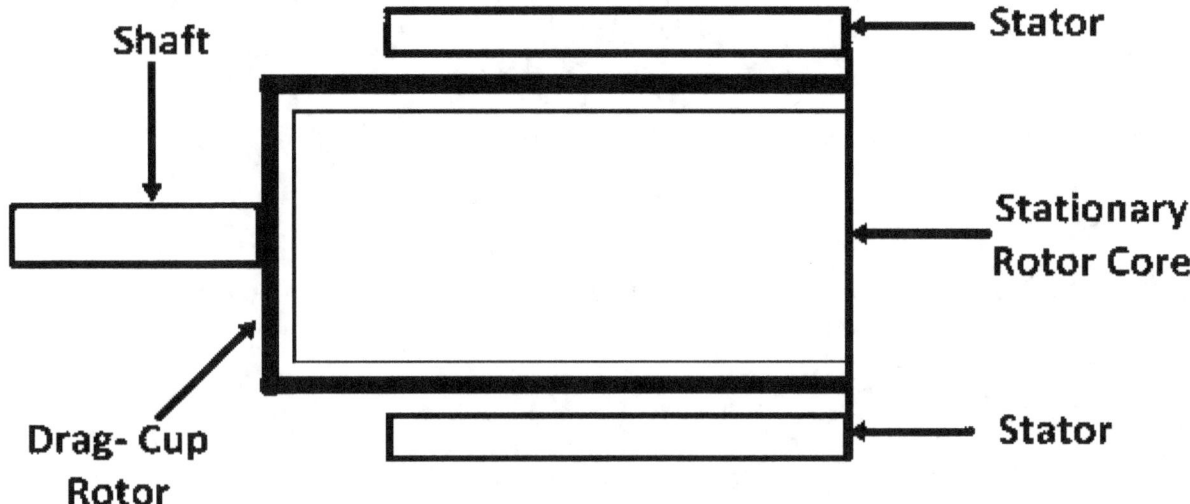

A thin cup of conductive non-magnetic substance serves as the rotor of the Drag cup servo motor. The conducting cup has a central iron core that is fixed in situ. The magnetic circuit is finished with this setup. Due to the thin material used to make the motor's rotor, it will have a high resistance, which causes a high starting torque.

Three Phase AC Servo motors

In high-power servo systems, three-phase induction motors with voltage control are commonly used as servo motors. However, conventional three-phase squirrel cage induction motors exhibit significant nonlinearity when linked to circuit devices. To address this issue, a control technique known as vector control or field-oriented control is employed to transform the motor into a linear decoupled machine. In this configuration, the current is adjusted to separate the torque and flux, resulting in a rapid increase in speed and torque.

Features

1. These gadgets are light in weight.
2. It provides both operational stability and dependability.
3. At the time of operation, hardly much noise is made.
4. It has torque-speed characteristics that are virtually linear.

5. The expense of maintenance is decreased because there aren't any brushes or slide rings here.

Use of AC Servomotors

Applications for AC servomotors include the following:

Due to the many benefits afforded by AC servomotors, these are primarily used in computers, position-controlling devices, and instruments that use servomechanism. Additionally, they are used in robotics machinery, machine tools, and tracking systems.

CHAPTER-15: APPLICATIONS OF THREE-PHASE INDUCTION MOTOR

A three-phase induction motor is also known as an electromechanical energy converter which converts the three-phase electrical input power into mechanical power at the output.

A 3-phase induction motor consists of a stator and a rotor. The stator is wound with three phases, whereas the rotor has a short-circuited winding known as the rotor winding. The stator winding is powered by the three-phase supply.

Working Principle of a 3-Phase Induction Motor

A part of a 3-phase induction motor may be used to demonstrate how it functions as follows:

When a balanced 3-phase supply is used to power the 3-phase stator winding, a rotating magnetic field (RMF) is produced in the motor. The synchronous speed at which this RMF revolves around the stator is determined by,

Synchronous speed $N_s = 120f / P$

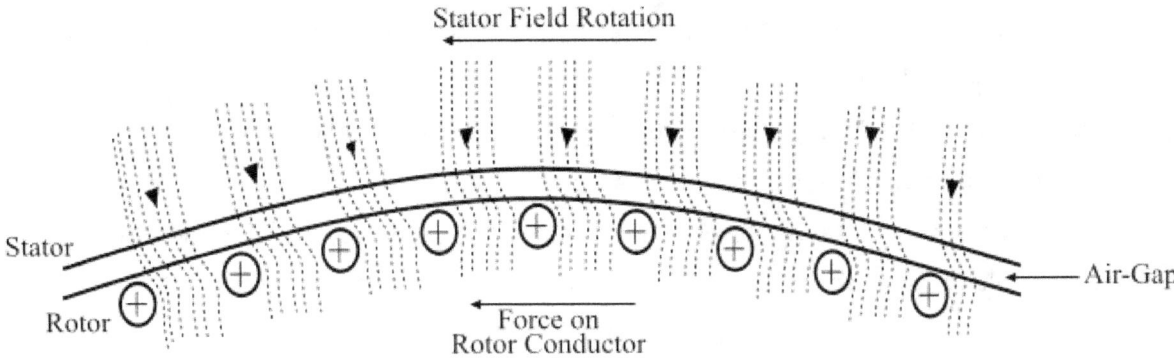

- The rotor conductors, which are still stationary, are severed by the RMF after it passes through the air gap. Due to the relative motion of the stationary rotor conductors and the RMF, EMFs are produced in the rotor conductors. Currents begin to flow in the rotor conductors as the rotor circuit is shorted out.
- Since the magnetic field generated by the stator winding is where the current-carrying rotor conductors are located. The rotor conductors therefore encounter mechanical force. The torque created by the combined mechanical forces acting on each conductor of the rotor causes it to move in the same direction as the revolving magnetic field. Consequently, a three-phase induction motor converts the three phases of electric power input into mechanical power at the output.
- In addition, the rotor should travel in the stator field's direction in accordance with Lenz's law, which states that rotor currents should generally run counter to the force that created them. The rotor currents in this scenario are due to the relative speed between the RMF and the rotor

conductors. To reduce this relative speed, the rotor starts to revolve in the same direction as the RMF.

Applications of 3-Phase Induction Motors

Three-phase induction motors with warped rotors or slip rings have a variety of uses.

1. Slip ring induction motors are appropriate for applications demanding low beginning current and loads requiring high starting torque.
2. When used with heavy loads, slip ring induction motors experience extremely large rotor energy losses during acceleration.
3. The application of slide ring induction motors extends to loads that must be gradually increased.
4. They are used when a load has to have its speed regulated.
5. A few examples of typical machinery that employ winding rotor or slip ring induction motors are crushers, plunger pumps, cranes & hoists, elevators, compressors, and conveyors.

Applications of 3-Phase Squirrel Cage Induction Motors

Induction motors with squirrel cages come in a variety of standard designs to meet the various starting and operating needs of various industrial applications. The most crucial design element for squirrel cage motors is the effective resistance of the rotor cage circuit.

The following list includes the uses for several kinds of squirrel cage induction motors:

Class A Motors

Normal beginning torque, high starting current, and minimal operating slip are all characteristics of class-A squirrel cage induction motors (from 0.005 to 0.015). This motor's single-cage rotor offers a low resistance. The efficiency of class-A motors is great at full load. As a result, these motors are appropriate for loads such as:

1. Blowers, Driers
2. Fans, Coolers
3. Machine tools, and

4. Centrifugal pumps etc.

Class B Motors

The beginning torque, starting current, and operating slip of Class-B motors are all normal. Utilizing a double-cage or deep bar rotor will preserve the beginning torque while decreasing the starting current by increasing the leakage reactance. The majority of people utilize these motors for full-voltage starting. As a result, they are also employed for loads like centrifugal pumps, fans, and blowers, among other things.

Class C Motors

High beginning torques and low starting current are characteristics of class-C motors. These motors use deep-bar rotors or twin cages with strong resistance. Class-C motors are used for basically constant-speed loads with moderately high torque and low beginning current requirements. Therefore, these motors are employed to move loads like **reciprocating pumps, conveyors, compressors, and crushers.**

Class D Motors

The class-D motors have the highest starting torque of any squirrel cage induction motors. These motors employ high resistance materials like brass for the rotor conductor bars instead of copper or aluminum. Low starting current and high operating slip (between 0.08 and 0.15), which result in low running efficiency, are characteristics of class-D squirrel cage motors. Therefore, these motors are utilized to power intermittent loads like punch presses, bulldozers, die-stamping machines, and shears that require heavy impact and quick acceleration. The motor is to be connected to a flywheel, which supplies the kinetic energy upon impact.

CHAPTER-16: WHAT IS DIELECTRIC HEATING?

Dielectric heating is the method of heating in which the dielectric materials are heated using a high-frequency alternating electric field, radio waves, or microwave electromagnetic radiation.

- Electronic heating, radio frequency (R.F.) heating, high frequency heating, and diathermy are other names for dielectric heating.
- Dielectric heating is mostly used to heat insulators such as wood, plastic, ceramics, etc. that cannot be heated quickly or evenly using conventional heating techniques.
- Dielectric heating requires an input supply with a frequency of between 10 and 15 MHz and an applied voltage of 20 kV.

Principle of Dielectric Heating

The capacitor's operating concept underlies the dielectric heating technique (i.e. electrostatic). As far as we are aware, dielectric heating is only effective when used to heat non-conducting materials. The dielectric material to be heated is sandwiched between two conducting electrodes in this heating technique, which uses alternating voltage at a high frequency.

The diagram below depicts the analogous circuit and schematic layout of the dielectric heating process.

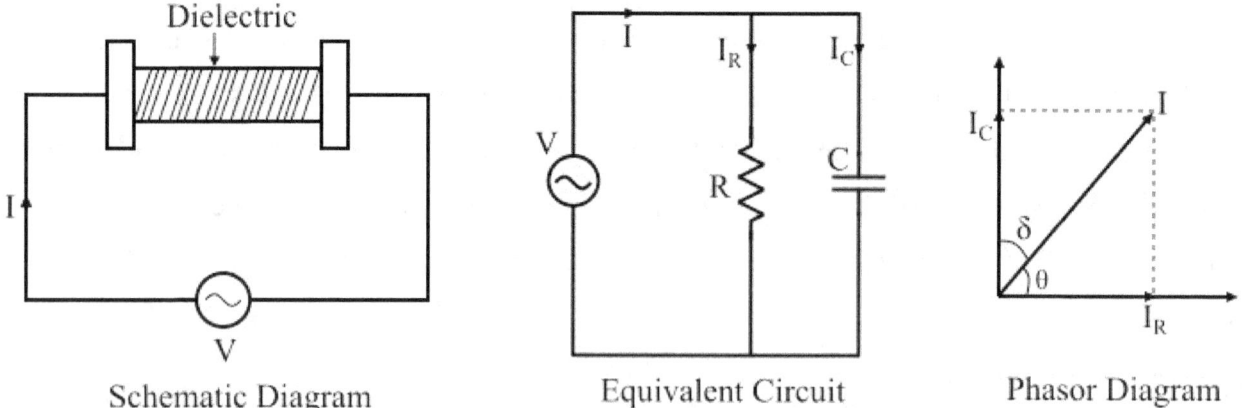

Schematic Diagram   Equivalent Circuit   Phasor Diagram

A high frequency potential difference is delivered across two electrodes that are separated from one another by a dielectric medium (the substance to be heated). This configuration creates a capacitor. The accompanying graphic also displays the corresponding circuit's phasor diagram. Due to the possibility that the capacitor created during the configuration for the dielectric heating is not pure, a resistor "R" is also presented in parallel in the circuit. However, the resistance, or "R," is quite high, resulting in a very little amount of current passing through it. Therefore, it may be said that the capacitor current Ic is equal to the entire current.

Applications of Dielectric Heating:

Dielectric heating is only utilized when other heating techniques are impractical since it is a costly heating technology. The following list includes some examples of dielectric heating applications:

Industry of Plastic

Numerous gadgets are made from plastic. Plastic molding will be quicker and error-free if the plastic is uniformly pre-heated. Before sending the plastic compounds into the molding area, the plastic is pre-heated using dielectric heating, which takes relatively little time to achieve the necessary heat in the plastic compounds.

Binding of books

The premium books are moved back and forth between tubular electrodes in alternating electrostatic fields. Through the use of dielectric heating, this dries the adhesive on the cover binding in a matter of seconds (about 5 to 10). Because of this, the book may be shipped and packaged without having to wait for the covers to dry. As a result, book binding uses dielectric heating.

Baking of sand cores

Before being employed in molding operations, the sand cores composed of dry sand, water, and some resin must be supported. These sand cores are dried by passing them through an alternate electrostatic field while being transported on a conveyor belt.

Industry of Plywood

The glue lines are dried using dielectric heating in the plywood industry. It is employed since, in normal circumstances, the glue lines may take a day or more to cure and fix the joint. However, it takes a few minutes for the joints to dry and solidify when they are heated using high frequency dielectric heating. Dielectric heating is essential for consistent heating in the plywood sector since the process of creating multilayer plywood requires quicker bonding.

Bakeries

For the backing of biscuits, cakes, and other bakery goods, the automatic bakeries' machines employ dielectric heating.

Digital sewing

Dielectric heating is also used while creating plastic apparel, including raincoats. If the raincoats were sewed on a standard sewing machine, rainwater may seep through the stitching gaps. In order to heat these materials more rapidly and evenly, cold rollers fed with high frequency supply (i.e., dielectric heating) are used.

Cigarette industry

Dielectric heating is being used in the tobacco industry to dry tobacco after it has been mixed with glycerin, etc.

Drying off oil emulsions

The moisture content of the oil emulsions is also taken out using dielectric heating. Because some chemical processes call for the blending of water and oil. The water content must then be eliminated in order to give the product the desired life.

Because the dielectric constants of oil and water are so different, the dielectric heating method is used for this purpose. As a result, when heating is accomplished through dielectric heating, water will boil off while temperature of oil will continue to stay significantly lower than the boiling point of water. The emulsion is securely, swiftly, and efficiently dried in this manner.

Food Dehydration

In order to dehydrate food so that it may be kept for a long time, the dielectric heating method is also used. Examples of food that has been dried through dielectric heating include dehydrated peas, packaged candies, etc. Through the use of an alternating electrostatic field, the food is put through a process that causes dielectric heating and, ultimately, dehydration.

Applications of Electromedicine

Dielectric heating is also employed in medicine to cure a variety of illnesses. Consider the procedure known as "diathermy," which involves heating the interior of the affected area using dielectric heating to treat a section of the body that is experiencing a particular form of discomfort.

The body portion is clamped by two electrodes during this procedure, and the electrodes are kept apart from the body. The body's tissues and bones between the electrodes produce dielectric heat as the electrodes are subsequently activated by a high frequency, high voltage power source.

www.ingramcontent.com/pod-product-compliance
Lightning Source LLC
Chambersburg PA
CBHW082209220526
45470CB00010B/3096